Food Chemistry

Food Chemistry

A Laboratory Manual

Second Edition

Dennis D. Miller
Cornell University
Ithaca, USA

C. K. Yeung
California Polytechnic State University
San Luis Obispo, USA

This edition first published 2022
© 2022 John Wiley & Sons, Inc.

Edition History
John Wiley & Sons, Inc. (1e, 1998)

The right of Dennis D. Miller and C. K. Yeung to be identified as the authors of this work has been asserted in accordance with law.

Registered Office
John Wiley & Sons, Inc., 111 River Street, Hoboken, NJ 07030, USA

Editorial Office
111 River Street, Hoboken, NJ 07030, USA

For details of our global editorial offices, customer services, and more information about Wiley products visit us at www.wiley.com.

Wiley also publishes its books in a variety of electronic formats and by print-on-demand. Some content that appears in standard print versions of this book may not be available in other formats.

Library of Congress Cataloging-in-Publication Data
Names: Miller, Dennis D., 1945– author. | Yeung, C. K., author.
Title: Food chemistry : a laboratory manual / Dennis D. Miller, Cornell
    University, Ithaka, USA, C. K. Yeung, California Polytechnic State
    University San Luis Obispo, USA .
Description: Second edition. | Hoboken, NJ : John Wiley & Sons, Inc., 2022.
    | Includes bibliographical references and index.
Identifiers: LCCN 2021034580 (print) | LCCN 2021034581 (ebook) | ISBN
    9780470639313 (paperback) | ISBN 9781119714583 (adobe pdf) | ISBN
    9781119714606 (epub)
Subjects: LCSH: Food–Analysis–Laboratory manuals. |
    Food–Composition–Laboratory manuals.
Classification: LCC TX541 .M49 2022 (print) | LCC TX541 (ebook) | DDC
    664/.07–dc23
LC record available at https://lccn.loc.gov/2021034580
LC ebook record available at https://lccn.loc.gov/2021034581

Cover images: © Rachan Panya/EyeEm/Getty Images, Chemical structure courtesy of Dennis D. Miller
Cover design by Wiley

Set in 9.5/12.5pt STIXTwoText by Straive, Pondicherry, India

SKY10033303_021422

Contents

Preface to the Second Edition   *xv*
Preface to the First Edition   *xvi*
Acknowledgments   *xvii*
About the Companion Website   *xix*

1   **Acids, Bases, and Buffers**   *1*
1.1   Learning Outcomes   *1*
1.2   Introduction   *1*
1.2.1   Acids   *1*
1.2.1.1 Food Acidulants   *2*
1.2.1.2 Reactions of Food Acids   *3*
1.2.2   Bases   *3*
1.2.3   Buffers   *3*
1.3   Apparatus and Instruments   *3*
1.4   Reagents and Materials   *4*
1.5   Procedures   *5*
1.5.1   Determining the pH of a Solid Food   *5*
1.5.2   Preparation of a Buffer and Determination of Buffer Capacity   *5*
1.6   Problem Set   *5*
1.7   References   *6*
1.8   Suggested Reading   *6*
         Answers to Problem Set   *6*

2   **Chemical Leavening Agents**   *7*
2.1   Learning Outcomes   *7*
2.2   Introduction   *7*
2.2.1   Chemical Leavening Agents   *8*
2.2.1.1 Baking Soda   *8*
2.2.1.2 Baking Powders   *8*
2.2.2   Neutralizing Values   *10*
2.2.3   Leavening Rates   *11*
2.2.4   Effect of Leavening Acid on Dough Rheology   *11*
2.3   Apparatus and Instruments   *11*

2.4   Reagents and Materials   *12*
2.5   Procedures   *12*
2.5.1   Determination of Leavening Rates   *12*
2.5.1.1 The Apparatus   *12*
2.5.1.2 Experimental Treatments and Controls   *12*
2.5.1.3 Protocol   *13*
2.5.1.4 Data Analysis   *13*
2.5.2   Chemically Leavened Biscuits   *13*
2.5.2.1 Biscuit Formula   *13*
2.5.2.2 Treatments   *14*
2.5.2.3 Protocol   *14*
2.5.2.4 Volume Determination of Biscuits   *14*
2.6   Problem Set   *14*
2.7   Useful Formulas and Values   *15*
2.8   References   *16*
2.9   Suggested Reading   *16*
         Answers to Problem Set   *16*

3   **Properties of Sugars**   *19*
3.1   Learning Outcomes   *19*
3.2   Introduction   *19*
3.3   Apparatus and Instruments   *20*
3.4   Reagents and Materials   *21*
3.5   Procedures   *21*
3.6   Study Questions   *22*
3.7   References   *22*
3.8   Suggested Reading   *22*

4   **Nonenzymatic Browning**   *23*
4.1   Learning Outcomes   *23*
4.2   Introduction   *23*
4.2.1   Caramelization   *23*
4.2.2   The Maillard Reaction   *24*
4.2.2.1 Sugar   *25*
4.2.2.2 Amine   *25*
4.2.2.3 Temperature   *26*
4.2.2.4 Concentration   *27*
4.2.2.5 pH   *27*
4.3   Apparatus and Instruments   *27*
4.4   Reagents and Materials   *28*
4.4.1   Reagents to Be Prepared by the Student   *28*
4.4.2   Reagents to Be Prepared by the Teaching Staff   *28*
4.5   Procedures   *28*
4.5.1   Preparation of a Glucose/Glycine Model System   *28*
4.5.2   Heating Experiment   *29*
4.5.3   Measurement of Extent of Browning   *29*

4.5.4   Browning in Nonfat Dry Milk (Demonstration)   *29*
4.5.5   Role of Milk in Crust Color of Bread (Demonstration)   *29*
4.5.6   Browning in Cookies   *30*
4.5.6.1 Sugar Cookie Formula   *30*
4.5.6.2 Baking Directions   *30*
4.6   Problem Set   *31*
4.7   Study Questions   *31*
4.8   References   *31*
4.9   Suggested Reading   *32*
         Answers to Problem Set   *32*

5   Food Hydrocolloids   *33*
5.1   Learning Outcomes   *33*
5.2   Introduction   *33*
5.2.1   Alginate   *34*
5.2.2   Alginate Gels   *35*
5.2.3   Carrageenan   *36*
5.2.4   Locust Bean Gum and Guar Gum   *37*
5.2.5   Xanthan Gum   *39*
5.3   Apparatus and Instruments   *39*
5.4   Reagents and Materials   *39*
5.5   Procedures   *40*
5.5.1   Effect of Heat Treatment on Gelation   *40*
5.5.2   Effect of Concentration on Viscosity   *40*
5.5.3   Emulsion Stability   *40*
5.5.4   Diffusion Setting and Internal Setting Alginate Gels   *41*
5.5.4.1 Diffusion Setting Gel   *41*
5.5.4.2 Internal Setting Gel   *41*
5.6   Study Questions   *41*
5.7   References   *41*
5.8   Suggested Reading   *42*

6   Functional Properties of Proteins   *43*
6.1   Learning Outcomes   *43*
6.2   Introduction   *43*
6.2.1   Soybean Processing: Soy Milk, Tofu, and Soybean Protein Isolate   *44*
6.2.2   Assaying Protein Concentration   *45*
6.3   Apparatus and Instruments   *45*
6.4   Reagents and Materials   *46*
6.5   Procedures   *46*
6.5.1   Standard Curve for the Bradford Protein Assay   *46*
6.5.2   Effect of pH on Protein Solubility   *46*
6.5.2.1 Preparation of Protein Extracts   *46*
6.5.2.2 Measurement of Protein Concentration in the Extracts   *47*
6.5.3   Preparation of Soy Protein Isolate and Tofu   *47*
6.5.3.1 Extraction   *47*

6.5.3.2   Soy Protein Isolation   *47*
6.5.3.3   Production of Tofu   *47*
6.6   Problem Set   *48*
6.7   Study Questions   *48*
6.8   References   *48*
6.9   Suggested Reading   *49*
          Answers to Problem Set   *49*

7   Lactose   *51*
7.1   Learning Outcomes   *51*
7.2   Introduction   *51*
7.2.1   Lactose Assay   *53*
7.3   Apparatus and Instruments   *54*
7.4   Reagents and Materials   *55*
7.5   Procedures   *55*
7.5.1   Lactose and D-galactose Assay Protocol   *55*
7.5.2   Lactase Assay   *55*
7.6   Experimental Design   *55*
7.7   Study Questions   *56*
7.8   References   *56*
7.9   Suggested Reading   *56*

8   Enzymatic Browning: Kinetics of Polyphenoloxidase   *57*
8.1   Learning Outcomes   *57*
8.2   Introduction   *57*
8.2.1   Enzyme Kinetics   *57*
8.2.2   PPO Assay   *61*
8.2.3   Control of Enzymatic Browning   *62*
8.3   Apparatus and Instruments   *62*
8.4   Reagents and Materials   *63*
8.5   Procedures   *63*
8.5.1   Preparation of Crude Enzyme Extract   *63*
8.5.2   Enzyme Assay   *63*
8.5.3   Data Treatment   *64*
8.5.4   Required Notebook Entries   *64*
8.6   Problem Set   *64*
8.7   Study Questions   *65*
8.8   References   *66*
          Answers to Problem Set   *66*

9   Blanching Effectiveness   *67*
9.1   Learning Outcomes   *67*
9.2   Introduction   *67*
9.3   Apparatus and Instruments   *69*
9.4   Reagents and Materials   *69*
9.5   Procedures   *70*
9.6   Study Questions   *70*

9.7   References   70
9.8   Suggested Reading   70

10   **Lipid Oxidation**   71
10.1   Learning Outcomes   71
10.2   Introduction   71
10.2.1 The Chemistry of Lipid Oxidation   71
10.2.2 Control of Lipid Oxidation   75
10.2.2.1 Elimination of Oxygen   75
10.2.2.2 Scavenging of Free Radicals   75
10.2.2.3 Chelation of Metal Ions   76
10.2.3 Measurement of Lipid Oxidation in Foods   76
10.2.3.1 Thiobarbituric Acid Test (TBA Test)   76
10.2.3.2 Peroxide Value   77
10.2.3.3 Conjugated Diene Methods   77
10.2.3.4 Oxygen Bomb Test   77
10.2.3.5 Total and Volatile Carbonyl Compounds   77
10.2.3.6 Anisidine Value Test   77
10.3   Apparatus and Instruments   78
10.4   Reagents and Materials   78
10.5   Procedures: Lipid Oxidation in Turkey Meat   78
10.6   Problem Set: Calculation of TBARS   79
10.7   Study Questions   80
10.8   References   81
10.9   Suggested Reading   81
        Answers to Problem Set   82

11   **Ascorbic Acid: Stability and Leachability**   83
11.1   Learning Outcomes   83
11.2   Introduction   83
11.2.1 Chemistry   83
11.2.2 Functions of Ascorbic Acid in Foods   85
11.2.2.1 Oxygen Scavenger   85
11.2.2.2 Free Radical Scavenger   86
11.2.2.3 Control of Enzymatic Browning   86
11.2.2.4 Dough Improver   87
11.2.3 Stability of Ascorbic Acid   87
11.2.4 Rationale for the Experiment   88
11.3   Apparatus and Instruments   88
11.4   Reagents and Materials   89
11.5   Procedures   89
11.5.1 Ascorbic Acid Standard Curve   89
11.5.2 Effect of pH on Ascorbic Acid Stability   89
11.5.3 Effects of Temperature, pH, and Cu^{2+} on the Stability of Ascorbic Acid   90
11.5.4 Effect of Cooking on the Ascorbic Acid Content of Cabbage   90
11.6   Problem Set   90
11.7   Study Questions   91

11.8   References   *91*
            Answers to Problem Set   *92*

12   **Hydrolytic Rancidity in Milk**   *93*
12.1   Learning Outcomes   *93*
12.2   Introduction   *93*
12.2.1   The Copper Soap Solvent Extraction Method   *94*
12.3   Apparatus and Instruments   *96*
12.4   Reagents and Materials   *96*
12.5   Treatments and Controls   *96*
12.6   Procedures   *97*
12.6.1   Standard Curve   *97*
12.6.2   Free Fatty Acids in Milk   *97*
12.6.3   Calculations   *97*
12.7   Problem Set   *98*
12.8   Study Questions   *98*
12.9   References   *98*
12.10   Suggested Reading   *98*
            Answers to Problem Set   *99*

13   **Caffeine in Beverages**   *101*
13.1   Learning Outcomes   *101*
13.2   Introduction   *101*
13.3   Apparatus and Instruments   *103*
13.4   Reagents and Materials   *103*
13.5   Operation of the HPLC   *103*
13.6   Procedures   *104*
13.6.1   Standard Curve   *104*
13.6.2   Caffeine in Soda and Energy Drinks   *105*
13.6.3   Caffeine in Coffee   *105*
13.6.4   Caffeine in Tea   *105*
13.7   Data Analysis   *105*
13.8   References   *105*
13.9   Suggested Reading   *106*

14   **Color Additives**   *107*
14.1   Learning Outcomes   *107*
14.2   Introduction   *107*
14.2.1   Binding to Wool   *110*
14.2.2   Removal from Wool   *110*
14.2.3   Solid-Phase Extraction (SPE)   *110*
14.2.4   Separation and Identification   *111*
14.3   Apparatus and Instruments   *111*
14.4   Reagents and Materials   *112*
14.5   Procedures   *112*
14.5.1   Qualitative Identification of Artificial Colors from Food Products   *112*
14.5.2   Separation and Identification of the Extracted Colors   *113*

14.5.3   Quantitative Analysis of FD&C Red Dye # 40 in Cranberry Juice   *113*
14.6   Study Questions   *114*
14.7   References   *114*
14.8   Suggested Reading   *114*

15   Plant Pigments   *115*
15.1   Learning Outcomes   *115*
15.2   Introduction   *115*
15.3   Apparatus and Instruments   *119*
15.4   Reagents and Materials   *119*
15.5   Procedures   *120*
15.5.1   Extraction and Separation of Lipid Soluble Plant Pigments   *120*
15.5.2   Extraction of Water Soluble Plant Pigments   *120*
15.5.3   Effect of pH on the Color of Water Soluble Plant Pigments   *120*
15.5.4   Demonstration   *121*
15.6   Study Questions   *121*
15.7   References   *121*
15.8   Suggested Reading   *121*

16   Meat Pigments   *123*
16.1   Learning Outcomes   *123*
16.2   Introduction   *123*
16.2.1   Meat Curing   *125*
16.2.2   Effect of Cooking on Meat Color   *126*
16.3   Apparatus and Instruments   *127*
16.4   Reagents and Materials   *127*
16.5   Procedures   *127*
16.5.1   Preparation and Spectral Analysis of Myoglobin, Oxymyoglobin,
           and Metmyoglobin   *127*
16.5.2   Preparation and Spectral Analysis of Nitric Oxide Myoglobin   *128*
16.5.3   Concentration of Metmyoglobin, Myoglobin, and Oxymyoglobin   *128*
16.5.4   Demonstration   *129*
16.6   Study Questions   *129*
16.7   References   *129*
16.8   Suggested Reading   *130*

17   Meat Tenderizers   *131*
17.1   Learning Outcomes   *131*
17.2   Introduction   *131*
17.3   Apparatus and Instruments   *132*
17.4   Reagents and Materials   *133*
17.5   Procedures   *133*
17.5.1   Preparation of Samples and Standards   *133*
17.5.1.1 Sample Treatments   *133*
17.5.1.2 Protein Extraction and Preparation for Electrophoresis   *134*
17.5.1.3 Preparation of SDS-PAGE Standards for Electrophoresis.   *134*
17.5.2   Electrophoresis   *134*

17.5.2.1 Loading and Running the Gel   *134*
17.5.2.2 Staining the Gel   *134*
17.5.3   Demonstration   *134*
17.6   Study Questions   *134*
17.7   References   *135*
17.8   Suggested Reading   *135*

18   Detection of Genetically Engineered Maize Varieties   *137*
18.1   Learning Outcomes   *137*
18.2   Introduction   *137*
18.2.1   Detection of a GE Protein by Immunoassay   *140*
18.2.2   Detection of a Trans Gene by PCR   *141*
18.3   Apparatus and Instruments   *143*
18.4   Reagents and Materials   *143*
18.5   Procedures   *144*
18.6   Study Questions   *145*
18.7   References   *145*
18.8   Suggested Reading   *146*

19   Food Emulsions and Surfactants   *147*
19.1   Learning Outcomes   *147*
19.2   Introduction   *147*
19.2.1   Emulsions   *147*
19.2.2   Surfactants   *147*
19.2.3   Surfactants in Food Systems   *148*
19.3   Part I – Butter Churning (Phase Inversion)   *150*
19.3.1   Materials and Methods   *150*
19.3.1.1 Materials for Buttermaking   *150*
19.3.1.2 Buttermaking Procedure   *150*
19.3.2   Study Questions   *151*
19.4   Part II – Margarine Manufacture (Use of Surfactant for Semi-solid Foods)   *151*
19.4.1   Materials and Methods   *151*
19.4.1.1 Materials for Margarine Manufacture   *151*
19.4.1.2 Manufacture Procedure   *152*
19.4.2   Study Questions   *152*
19.5   Part III – Dispersion of Eugenol in Water (Surfactant Solubilization Capacity)   *152*
19.5.1   Materials and Methods   *153*
19.5.1.1 Materials for Dispersion Experiment   *153*
19.5.1.2 Experimental Procedure   *153*
19.5.2   Study Questions   *154*
19.6   Part IV – Mayonnaise Stability   *155*
19.6.1   Materials and Methods   *155*
19.6.1.1 Materials for Mayonnaise Experiment   *155*
19.6.1.2 Experimental Procedure   *155*
19.6.2   Study Questions   *156*
19.7   References   *156*
19.8   Suggested Reading   *158*

Appendix I   *159*
Conversion Factors   *159*

Appendix II   *161*
Concentration   *161*
Definition   *161*
Suggested Reading   *162*

Appendix III   *163*
Acids, Bases, Buffers, and pH Measurement   *163*
Review of pH and Acid–Base Equilibria   *163*
Acids and Bases   *163*
Acid/Base Equilibria   *163*
The pH Scale   *165*
pK   *165*
Buffers: Functions and Uses   *166*
Problems   *167*
Choosing a Buffer System   *169*
Preparation of Buffers   *171*
Activity and Ionic Strength   *173*
pH Measurement   *174*
Making pH Measurements   *175*
References   *176*
Suggested Reading   *176*

Appendix IV   *177*
Spectrophotometry   *177*
Introduction   *177*
Operation of a Spectrophotometer   *180*
Notes for Operators   *180*
Problem Set   *180*
References   *181*
Answers to Problem Set   *181*

Appendix V   *183*
Chromatography   *183*
What Is Chromatography?   *183*
Chromatography Terminology   *183*
Types of Chromatography   *184*
Adsorption Chromatography (AC)   *185*
Liquid–Liquid Partition Chromatography (LLPC)   *185*
Bonded Phase Chromatography (BPC)   *185*
Ion-Exchange Chromatography (IEC)   *185*
Gel Permeation Chromatography (GPC)   *185*
High-Performance Liquid Chromatography   *186*
The HPLC System   *187*
References   *188*
Suggested Reading   *189*

Appendix VI   *191*
Electrophoresis   *191*
Introduction   *191*
References   *195*
Suggested Reading   *196*

Appendix VII   *197*
Glossary   *197*

Preface to the Second Edition

Food Chemistry: A Laboratory Manual, first edition, has been adopted by dozens of universities in the United States and internationally since it was published in 1998. The second edition has been extensively revised and with new chapters added. I was extremely fortunate to have Professor C.K. (Vincent) Yeung join me as a coauthor for the second edition. Dr. Yeung holds a B.S. in chemistry from the Chinese University of Hong Kong, an M.S. in dairy products technology from California Polytechnic State University, and a Ph.D. in food science and technology from Cornell University. He is currently Associate Professor of Dairy Science at California Polytechnic State University in San Luis Obispo, California. Dr. Yeung's knowledge and insights gained from his strong educational background and his years of teaching dairy chemistry laboratory classes were invaluable in making improvements in existing exercises and in developing new ones.

Our overarching goal in designing the laboratory exercises in the manual was to help students develop an in-depth understanding of the fundamental chemical principles that underlie relationships between the composition of foods and food ingredients and their functional, nutritional, and sensory properties. In addition, students who complete the laboratory exercises will learn and practice many methods and techniques common in food chemistry research and food product development. We recommend the manual for a 2-credit food chemistry laboratory course although it contains many more exercises (19) than can be reasonably completed in a 1-semester course. This should allow instructors to select exercises that most closely provide the learning outcomes they wish to achieve.

Each chapter includes introductory summaries of key concepts and principles that are important for understanding the methods used and interpreting the results obtained from the experiments. In writing and revising these summaries, we relied heavily on two widely adopted food chemistry textbooks: *Introductory Food Chemistry* by John W. Brady, Cornell University Press, 2013, and *Fennema's Food Chemistry*, 5th edition, S. Damodaran and K. L. Parkin, editors, CRC Press Taylor & Francis Group, 2017. We encourage students to read relevant sections in these and/or other food chemistry textbooks in addition to the introductory material in the manual for a more rigorous discussion of the relevant topics.

Dennis D. Miller
Professor Emeritus
Department of Food Science, Cornell University
October 2021

Preface to the First Edition

Food chemistry is a broad discipline that draws on principles of physical, organic, and biological chemistry. Advances in food chemistry over the past century have had a dramatic impact on our understanding of all aspects of food science and technology and have played a major role in the improvement of the quality, quantity, and availability of the food supply.

This manual is designed for a one-semester laboratory course in food chemistry. Emphasis is placed on understanding fundamental chemical principles that underlie relationships between the composition of foods and functional, nutritional, and organoleptic properties. In addition, many laboratory techniques that are common in basic and applied research in food chemistry are introduced.

Students should have a background in general, organic, and biochemistry as well as a concurrent lecture course in food chemistry. Each experiment is preceded by an introduction of the principles necessary for understanding and interpreting the data. Students are encouraged to supplement the introductory material by reading selected sections in a comprehensive food chemistry textbook in addition to the references cited at the end of each experiment.

Many students have successfully performed the experiments described in the manual. I have tried diligently to eliminate errors and provide clear instructions for students and instructors. Nevertheless, errors and unclear writing have a way of creeping in. I would appreciate hearing from students and instructors when they find errors.

Dennis D. Miller
Ithaca, New York

Acknowledgments

This manual is the product of the efforts of many people over a period of more than 40 years. First and foremost, we thank the students at Cornell University who enrolled in the Food Chemistry Laboratory course in the Department of Food Science over the 40 years it was offered as part of the Institute of Food Technologists-approved curriculum for Food Science majors. Their enthusiasm, lab reports, presentations, questions, and gentle criticisms were an immense help in developing and refining the experiments in the manual. Another group of students who contributed enormously were the graduate and undergraduate teaching assistants who set up the laboratories, assisted students during the lab periods, graded lab reports, and offered suggestions for improving the exercises. In recent years, Aaron Jacobsen, Teaching Support Specialist at Cornell, has worked tirelessly to streamline the lab exercises and identify errors in the procedures. We also thank our colleagues at Cornell and Cal Poly for their comments, suggestions, and continual intellectual stimulation. We thank Drs. Alicia Orta Ramirez and Motoko Mukai who both took their turns in teaching the Food Chemistry Lab course at Cornell. We are especially grateful to Professors John Brady and Chang Lee for their friendship, encouragement, and support. Last but not least, we thank our editors at Wiley for their expert advice and support.

About the Companion Website

This book is accompanied by a companion website.

www.wiley.com/go/Miller/foodchemistry2

This website includes:

- **Preparations and solutions for all chapters**

1

Acids, Bases, and Buffers

1.1 Learning Outcomes

After completing this exercise, students will be able to:

1) Explain the roles of acids and bases in food products.
2) Measure the pH of selected food products.
3) Prepare and evaluate a buffer system.
4) Measure the buffering capacity of a common beverage.

1.2 Introduction

Many food components may be classified as acids or bases due to their capacity to donate or accept protons (hydrogen ions). These components perform numerous important functions including flavor enhancement, control of microbial growth, inhibition of browning, alteration of texture, prevention of lipid oxidation, and pH control.

Acids and bases are key metabolites in living plant and animal organisms, for example as intermediates in the TCA cycle, and are mostly retained when the plant is harvested or the animal is slaughtered so they are naturally present in foods. They may also be added during processing or synthesized during fermentation to produce desired characteristics in the final food product.

The concentration and relative proportion of acids and bases determine the pH of a food, an extremely important characteristic. pH can affect the flavor, color, texture, stability, and behavior in food processing situations. For example, commercial sterilization of acid foods (pH less than 4.6) [1] can be achieved under milder processing conditions than in foods with a higher pH.

1.2.1 Acids

Acids serve a variety of functions in foods including flavor enhancement, control of microbial growth, protein coagulation, emulsification, control of browning, buffering action, and metal chelation (to control lipid oxidation). All acids have a sour taste but different acids produce

Food Chemistry: A Laboratory Manual, Second Edition. Dennis D. Miller and C. K. Yeung.
© 2022 John Wiley & Sons, Inc. Published 2022 by John Wiley & Sons, Inc.
Companion website: www.wiley.com/go/Miller/foodchemistry2

Table 1.1 Acids common in foods: structures and pK_a values.

Substance	Structure	pK_a	Food found in
Acetic acid	CH_3COOH	$pK = 4.75$	Vinegar, figs
Adipic acid	$HOOC(CH_2)_4COOH$	$pK_1 = 4.43$ $pK_2 = 5.62$	Beets
Butyric acid	$CH_3(CH_2)_2COOH$	$pK = 4.82$	Cheese, butter
Citric acid	CH₂COOH | HO—C—COOH | CH₂COOH	$pK_1 = 3.06$ $pK_2 = 4.74$ $pK_3 = 5.40$	Oranges, lemons, apricots, tomatoes
Lactic acid	CH₃CHCOOH | OH	$pK = 3.83$	Yogurt, buttermilk, cheese, beer
Malic acid	HOOCCH₂CHCOOH | OH	$pK_1 = 3.40$ $pK_2 = 5.05$	Apples, apricots, grapes, oranges, tomatoes
Oxalic acid	COOH | COOH	$pK_1 = 1.27$ $pK_2 = 4.27$	Spinach, potatoes, tomatoes
Phosphoric acid	OH | O=P—OH | OH	$pK_1 = 2.12$ $pK_2 = 7.21$ $pK_3 = 12.32$	Tomatoes, acidulant used in soft drinks
Tartaric acid	OH | HOOCCHCHCOOH | OH	$pK_1 = 2.98$ $pK_2 = 4.34$	Grapes
Sodium hydrogen sulfate or sodium acid sulfate	$HO-SO_2-O^- Na^+$	$pK = 1.99$	Acidulant. Lowers pH without imparting acid taste. May be added to process water to enhance chlorine activity

distinctively different sour flavors. Thus, it is not enough to simply add any acid when attempting to produce a characteristic sour flavor in a food. Table 1.1 gives structures and pK values of some common food acids.

1.2.1.1 Food Acidulants
In the food industry, food additives that have acidic properties are commonly known as food acidulants. There are many approved food acidulants, but only a few are in wide use. They include organic acids like acetic acid, citric acid, fumaric acid, lactic acid, malic acid, and tartaric acid as well as the mineral acids phosphoric acid and sodium hydrogen sulfate. (See [2] for guidance in selecting food acidulants.)

1.2.1.2 Reactions of Food Acids

Most naturally occurring food acids are carboxylic acids. Carboxylic acids are weak acids compared with mineral acids such as HCl and H_2SO_4. Important reactions of carboxylic acids include the following:

Ionization:

Acetic acid   Acetate ion

Reaction with alcohols to form esters:

Acetic acid   Methanol   Methyl acetate

1.2.2 Bases

Bases are also common food additives and are added for a variety of purposes. They may be added to modify the flavor, color, and texture, enhance browning, induce chemical peeling, and produce CO_2. Examples of bases used as food additives include dilute NaOH (to induce chemical peeling in fruits and vegetables, enhance browning, de-bitter olives, solubilize proteins), phosphate salts (to prevent protein coagulation in evaporated and condensed milks, produce a smooth texture in processed cheese), and $NaHCO_3$ (to give chocolate a darker color, produce CO_2 in leavening systems).

1.2.3 Buffers

Buffers stabilize the pH in foods. They are also used to neutralize foods which are too acidic. By using the salt of the acid already present, acidity is reduced without adding neutralization flavors. Many buffers are present naturally in foods. Animal products are usually buffered by amino acids, proteins, and phosphate salts. In plants, organic acids (such as citric, malic, oxalic, and tartaric) in conjunction with phosphate salts are the primary buffers. Table 1.2 shows the pHs of some common foods. Notice that most foods are buffered in the acidic range (pH < 7).

See Appendix III or your chemistry and biochemistry textbooks for a review of acid and base chemistry.

1.3 Apparatus and Instruments

1) pH meter equipped with a pH electrode
2) Analytical balance
3) Household blender
4) Centrifuge
5) Centrifuge tubes

Table 1.2 Approximate pH values for some common foods[a].

Food	pH	Food	pH
Lime juice	2.0	Yogurt	4.0–4.5
Lemon juice	2.2	Cheddar cheese	5.1–5.5
Vinegar	2.6	Beef, fresh	5.5–5.0
Rhubarb	3.0	Pork, fresh	5.6–6.9
Grape juice	3.1–3.2	Turkey, fresh	5.7–6.1
Wines	2.9–3.9	Tuna	6.0
Apple juice	3.5–3.9	Carrots, fresh	5.7–6.1
Strawberries	3.2–3.4	Potatoes, fresh	6.1
Peaches	3.8	Green beans, fresh	6.5–6.7
Pears	3.9	Milk, fresh	6.6
Grapefruit juice	4.0	Sweet corn, fresh	6.7
Orange juice	4.2	Egg yolk	6.0–6.9
Tomato juice	3.8–4.7	Egg white (pH increases as egg ages)	7.6–9.2

[a] Modified from [3] and [4].

6) Pipette and pipette bulb, 10 ml
7) Volumetric flask, 200 ml
8) Beakers, 150 ml
9) Burette, 25 or 50 ml
10) Burette holder and stand
11) Thermometer
12) Funnel
13) Graduated cylinder, 100 ml
14) Squeeze bottle for deionized water
15) Tissue
16) Weighing paper
17) Spatula
18) Stirring hot plate with stirring bars

1.4 Reagents and Materials

1) Citric acid, monohydrate. MW = 210 g mole^{-1}
2) KOH, 0.5 N
3) HCl, 0.5 N
4) HCl, 0.001 N
5) Sprite® (Coca Cola Company) or comparable lemon-flavored soda
6) Selected vegetables, e.g. fresh and canned tomatoes
7) Calibration buffers, pH 2 and 4

1.5 Procedures

1.5.1 Determining the pH of a Solid Food [5]

1) Cut a fresh tomato into small cubes and blend in a blender until a uniform slurry is formed, measure the temperature of the slurry.
2) Calibrate your pH meter.
3) Measure the pH of the slurry.
4) Centrifuge an aliquot of the slurry for 10 minutes at maximum speed.
5) Measure the pH of the supernatant.
6) Repeat Steps 1 through 5 using canned tomatoes.

1.5.2 Preparation of a Buffer and Determination of Buffer Capacity

1) Calculate the amounts of citric acid monohydrate and 0.5 N KOH required to prepare 200 ml of 0.05 M citrate buffer, pH 3. **Note**: Even though citric acid is a triprotic acid, calculations for this pH range are made using $pK_a = 3.06$.
2) Prepare 200 ml of the buffer.
3) Measure the pH of your buffer. Is it 3.0? If not, can you explain why?
4) Determine the buffer capacity of your buffer in the alkaline direction by titrating a 100 ml aliquot with 0.5 N KOH. Express buffer capacity as the number of moles of OH^- required to increase the pH of 1 l of the buffer by 1 pH unit.
5) Repeat Step 4 using 0.001 N HCl in place of the citrate buffer, i.e. determine the buffer capacity of 0.001 N HCl.
6) Determine the buffer capacity of your buffer and 0.001 N HCl by calculation. Compare your experimental results with your calculated answers. Explain any discrepancies between experimental and calculated values.
7) Determine the buffer capacity of Sprite® in the same way you did for your citrate buffer (Step 4 above).

1.6 Problem Set

1 The K_a for the weak acid HA is 4.0×10^{-6}. What is the pH of a 0.01 M solution of this acid? What is its pK_a?

2 How many grams of acetic acid and sodium acetate are required to prepare 1.0 l of 0.5 M acetate buffer, pH 4.5? The pK_a for acetic acid is 4.75.

3 Explain carefully how to prepare 1.0 l of 0.05 M phosphate buffer, pH 7.0 from $NaH_2PO_4{\cdot}H_2O$ and 1.0 N NaOH or 1.0 N HCl. The molecular weight of sodium phosphate monohydrate monobasic is 138 g mole^{-1}. The pK_{a2} for H_3PO_4 is 7.21. **Hint**: To determine whether you will need to add NaOH or HCl, you need to calculate the pH of a 0.05 M solution of NaH_2PO_4.

4 Preparation of buffers using published tables.
   a) Using Tables III.2a, III.2b, and III.2c in Appendix III, describe carefully how you would prepare 1 l of 0.1 M acetate buffer, pH 5.2 and 1 l of 1/15 M phosphate buffer, pH 7.6.

b) Using the Henderson–Hasselbalch equation (shown below), calculate the theoretical pH of these 2 buffers. Explain why the calculated pHs are not exactly the same as the pHs shown in the Appendix table.

$$pH = pK_a + \log\frac{\left[A^-\right]}{\left[HA\right]}$$

1.7 References

1 FDA (2019). Acidified & low-acid canned foods guidance documents & regulatory information [Internet]. FDA [cited 2020 Mar 5]. http://www.fda.gov/food/guidance-documents-regulatory-information-topic-food-and-dietary-supplements/acidified-low-acid-canned-foods-guidance-documents-regulatory-information (accessed 5 March 2020).
2 Bartek Ingredients, Inc (2020). Self teaching guide for food acidulants - Google search [Internet]. [cited 2020 Mar 5] p. 37. https://www.google.com/search?q=Self+teaching+guide+for+food+acidulants&rlz=1C1EJFC_enUS825US876&oq=Self+teaching+guide+for+food+acidulants&aqs=chrome..69i57j69i60.2887j0j4&sourceid=chrome&ie=UTF-8 (accessed 5 March 2020).
3 USDA ARS (2020). pH of selected foods [Internet]. [cited 2020 Mar 5]. https://pmp.errc.ars.usda.gov/phOfSelectedFoods.aspx (accessed 5 March 2020).
4 Bennion, M. (1980). *The Science of Food*. San Francisco: Harper & Row. 598 p.
5 AOAC Official Method 981.12 (1982). *pH of Acidified Foods*. AOAC International.

1.8 Suggested Reading

Lindsay, R.C. (2017). Food additives. In: *Fennema's Food Chemistry*, 5e (eds. S. Damodaran and K.L. Parkin), 803–864. Boca Raton: CRC Press, Taylor & Francis Group.
Segel, I.H. (1976). *Biochemical Calculations: How to Solve Mathematical Problems in General Biochemistry*, 2e. New York: Wiley. 441 p.

Answers to Problem Set

1 pH = 3.7; pK_a = 5.4.
2 14.68 g sodium acetate; 19.26 g acetic acid.
3 Use NaOH to adjust pH; dissolve 6 g NaH_2PO_4 (or 6.9 g $NaH_2PO_4 \cdot H_2O$) in water (~900 ml); using pH meter, titrate to pH 7.0 with 1.0 N NaOH; transfer to a volumetric flask and dilute to 1 l.
4 a) pH 5.2 acetate buffer: mix 768 ml 0.1 M sodium acetate and 232 ml acetic acid. pH 7.6 phosphate buffer: mix 128 ml 1/15 M KH_2PO_4 and 872 ml 1/15 M Na_2HPO_4.
   b) Calculated pHs: Acetate buffer = 5.27; Phosphate buffer = 8.04.

2

Chemical Leavening Agents

2.1 Learning Outcomes

After completing this exercise, students will be able to:

1) Write balanced chemical equations showing the production of CO_2 in various chemical leavening systems.
2) Determine, experimentally, the rates of CO_2 release from selected leavening systems.
3) Select a chemical leavening system suitable for making biscuits, muffins, and other baked products.
4) Use the ideal gas law to calculate the volume of available CO_2 in a leavening system.
5) Calculate the weight of a given leavening acid required to neutralize a given amount of $NaHCO_3$.

2.2 Introduction

Leavening is derived from the Latin word *levo* which means raising or making light. Batters and doughs containing wheat or rye flour may be leavened by incorporating gases into them. Five gases, either alone or in combination, may be used for leavening: carbon dioxide, water vapor, ethanol vapor, air, and ammonia.

The incorporation of leavening gas into baked goods is most frequently accomplished by either yeast fermentation or chemical leavening:

1) Yeast fermentation of sugars:

$$C_6H_{12}O_6 \xrightarrow{\text{yeast}} 2CO_2 + 2C_2H_5OH \tag{1}$$

2) Chemical leavening:
   a) Decomposition of salts:

$$NH_4HCO_3 \xrightarrow{\text{heat}} NH_3 + H_2O + CO_2 \tag{2}$$

   b) Reaction of acids or acidic salts (HA) with sodium bicarbonate:

$$HA + NaHCO_3 \xrightarrow{\text{H}_2\text{O+heat}} NaA + H_2O + CO_2 \tag{3}$$

Food Chemistry: A Laboratory Manual, Second Edition. Dennis D. Miller and C. K. Yeung.
© 2022 John Wiley & Sons, Inc. Published 2022 by John Wiley & Sons, Inc.
Companion website: www.wiley.com/go/Miller/foodchemistry2

In yeast leavened products, the CO_2 is produced slowly, inflating air bubbles that have been previously incorporated into the dough during mixing. In contrast, chemical leavening provides some of the initial nucleating gas and provides a much faster means of inflating existing air bubbles during mixing and baking. When heated, the expanded dough "sets" forming the desired texture and lightness. Although seemingly simple in principle, a basic knowledge of chemical leavening is necessary in order to obtain the correct rate and amount of CO_2 release for a specific application.

2.2.1 Chemical Leavening Agents

2.2.1.1 Baking Soda

Baking soda (also known as bicarbonate of soda) is pure sodium bicarbonate ($NaHCO_3$). It releases CO_2 on reaction with acids (Eq. 3). It is commonly used in batters for muffins, pancakes, and cookies, in combination with a baking powder (see below). The source of acid in these batters may be sour milk (lactic acid), cultured buttermilk (lactic acid), molasses (various organic acids including acetic, propionic, and aconitic), lemon juice (citric acid), etc.

2.2.1.2 Baking Powders

Generally, baking powders contain three materials: (i) a CO_2 source, (ii) one or more leavening acids, and (iii) a diluent or filler.

CO_2 Source: Sodium bicarbonate is the most commonly used source of CO_2 (Eq. 3). Ammonium bicarbonate is sometimes used in cookies and crackers; however, care must be taken to remove any residual NH_3 gas formed during decomposition in order to alleviate off-flavors (Eq. 2). Sodium bicarbonate dissolves almost instantly but requires some form of acid to release the CO_2 (Eq. 3). Thus, the rate and degree of dissolution of the acid in baking powders governs the rate of CO_2 release. Thermal decomposition of sodium bicarbonate does not occur to any large extent, except under conditions of excess soda and/or very high temperatures.

Leavening Acids: Leavening acids that react with bicarbonate to release CO_2 rapidly when water or milk is added are called "fast-acting." "Slow-acting" acids react more slowly because they dissolve slowly at room temperature. Little CO_2 is liberated by slow-acting acids until the system is heated. Heating increases the rate of dissolution thereby increasing the rate of CO_2 generation [1]. A description of some common leavening acids used in baking powders follows:

i) *Monocalcium phosphate monohydrate* (MCP-H_2O): $Ca(H_2PO_4)_2 \cdot H_2O$. This leavening acid is used in most household baking powders. It is "fast-acting." In solution, it dissociates to form Ca^{2+} and $H_2PO_4^-$. The pK_a of $H_2PO_4^-$ is 7.21 while the pK_a of HCO_3^- is 10.33. Thus, $H_2PO_4^-$ is a much stronger acid than HCO_3^- and readily donates a proton to HCO_3^- which, in this case, acts as a base:

$$H_2PO_4^- + HCO_3^- \rightarrow HPO_4^{2-} + H_2O + CO_2 \tag{4}$$

One advantage of monocalcium phosphate over some other leavening acids is that it does not contain sodium.

ii) *Sodium Aluminum Sulfate* (SAS): $Na_2SO_4 \cdot Al_2(SO_4)_3$. SAS is frequently used in combination with MCP-H_2O in household baking powders to produce a so-called "double-acting" baking powder. SAS releases CO_2 from soda only at elevated temperatures. Excess SAS may adversely affect the flour gluten causing finished products to have a dull color and slightly bitter taste [1]. Acid is produced when $Al_2(SO_4)_3$ reacts with water to form sulfuric acid:

$$Al_2(SO_4)_3 + 6H_2O \rightarrow 2Al(OH)_3 + 3H_2SO_4 \qquad (5)$$

Once formed, sulfuric acid, a strong mineral acid, reacts rapidly with bicarbonate to generate CO_2.

iii) *Potassium Hydrogen Tartrate (Cream of Tartar):* $KHC_4H_4O_6$. Cream of tartar, a fast leavening acid, is not widely used in commercial baking powder formulations. It may be used in angel food cakes. Tartaric acid itself ($H_2C_4H_4O_6$) is a fast-acting acid and may be used in combination with the tartrate salt.

iv) *Coated Anhydrous Monocalcium Phosphate.* Coating MCP with slightly soluble compounds converts it to a slow-acting acid. Coatings may include slightly soluble calcium and aluminum phosphates.

v) *Glucono-delta-lactone* (GDL). This is a slow-acting acid and is stable in refrigerated and frozen doughs. Lactones are cyclic esters formed when alcohol and carboxylic acid groups on the same molecule react. GDL hydrolyzes slowly in water to yield gluconic acid.

$$(6)$$

Glucono-delta-lactone   Gluconic acid

vi) *Sodium Acid Pyrophosphate* (SAPP): $Na_2H_2P_2O_7$. SAPP is available in several grades produced by varying manufacturing conditions. The different grades are identified by numbers which are related to rates of reaction with sodium bicarbonate at room temperature. For example, SAPP-21, SAPP-28, and SAPP-40 represent grades with increasing reaction rates. SAPP-21 is used when there is a delay between mixing and baking, for example in refrigerated biscuits [2].

vii) *Sodium Aluminum Phosphate* (SALP). $NaAl_3H_{14}(PO_4)_8 \cdot 4H_2O$. A slow-acting acid, SALP has replaced by SAPP in some applications because SAPP residual salts may impart an off-flavor.

viii) *Other Food Ingredients.* Sour milk, cultured buttermilk, molasses, and ionic flour proteins are all acidic and therefore can act as leavening acids.

Diluents: Diluents are added to baking powders to prevent premature reaction of the sodium bicarbonate and acid, i.e. to provide physical separation of soda and acid, and to increase the bulk which makes measuring small quantities of sodium bicarbonate and leavening acid easier. Corn starch is the most common diluent. FDA requires that baking powders yield at least 12 g of CO_2 for every 100 g of powder. In order to achieve this, most formulations contain 26–30% sodium bicarbonate [3–5].

2.2.2 Neutralizing Values

Neutralizing values (NV) of leavening acids are used to compare the available acidity of the acids and to calculate the amount of acid needed in a given formulation of chemically leavened product. Neutralizing value is defined as the weight of $NaHCO_3$ that can be neutralized by 100 parts of the leavening acid. For example, if 100 g of an acid will neutralize 50 g of $NaHCO_3$, the neutralizing value of the acid is 50. NV are determined by titrating to a specific pH endpoint [6]. The formula for NV is:

$$NV = \frac{b}{a} \times 100 \tag{7}$$

where b = wt of $NaHCO_3$ neutralized;
a = wt of acid required to neutralize b.

See Table 2.1 for NV for some common leavening acids.

NV are used to determine the amount of acid needed to neutralize a given amount of $NaHCO_3$. For example, if a formula specifies 20 kg of $NaHCO_3$, the amount of SAS (NV = 104) required would be:

$$a = \frac{b}{NV} \times 100 = \frac{20\,kg}{104} \times 100 = 19\,kg \tag{8}$$

Baking powders and recipes are generally formulated to give baked products with near-neutral pHs. There are, however, some exceptions to this:

i) *Devil's Food Chocolate Cake.* Excess sodium bicarbonate is added to give an alkaline pH in order for the chocolate to form the characteristic deep red color.
ii) *Buttermilk products.* Excess acid (in the form of buttermilk) is added to give the characteristic flavor.
iii) *Pretzels and Gingerbread.* Alkaline pHs are produced to accelerate nonenzymatic browning reactions which are important for the deep brown color.

Table 2.1 Neutralizing values for some common leavening acids[a].

Leavening acid	Molecular weight	Neutralizing value	Reaction rate[b]
Monocalcium phosphate monohydrate (MCP)	252	80	Fast
Delayed anhydrous monocalcium phosphate (AMCP)	234	82	Slow
Sodium acid pyrophosphate (SAPP)	222	72	Slow
Sodium aluminum phosphate (SALP)	950	100	Slow
Dicalcium phosphate dihydrate (DCP)	172	33	Slow
Sodium aluminum sulfate (SAS)	484	104	Slow
Fumaric acid	116	145	Fast
Potassium acid tartrate	188	50	Medium
Glucono-delta-lactone	178	45	Slow

[a] Adapted from [4] and [7].
[b] Rate of reaction with $NaHCO_3$ at room temperature.

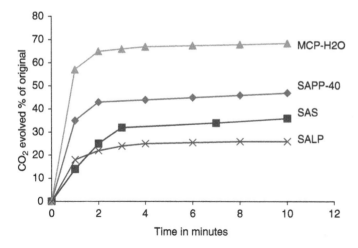

Figure 2.1 Rates of CO_2 production from mixtures of SALP and MCP-H_2O, SAPP-40, or SAS. Evolved CO_2 is expressed as a percentage of "available" CO_2 in the $NaHCO_3$. *Source:* Adapted from [9].

2.2.3 Leavening Rates

The rates at which doughs are leavened are important determinants of the quality of baked products. Leavening rates in dough systems are influenced by a variety of factors including the types and concentrations of the leavening agents, temperature, the availability of water, and pressure [8].

Leavening rates may be measured by trapping evolved CO_2 and plotting CO_2 volume versus time. The amount of CO_2 produced is usually expressed as a percentage of the total amount of CO_2 that would be released if all of the sodium bicarbonate were converted to CO_2 and H_2O. When measured in a dough system, leavening rates are called "dough reaction rates." Typical dough reaction rates for some leavening acids are shown in Figure 2.1.

Measurement of leavening rates must be carefully standardized with respect to temperature, time, agitation, and ingredients so that comparisons between laboratories and experiments will be valid [10]. The examples of leavening rates given above serve to illustrate the fact that rates differ considerably among leavening systems. It should be remembered that leavening rates for a given leavening acid can be manipulated by applying coatings and changing particle size. Also, other ingredients in the batter or dough may affect leavening rates. Thus, when purchasing leavening acids from suppliers, it is important to specify the leavening rate needed for a particular application.

2.2.4 Effect of Leavening Acid on Dough Rheology

Cations and anions from chemical leaveners may alter the elastic and viscous properties of the dough and the texture and resiliency of the crumb. Calcium and aluminum ions can prevent coalescence of air bubbles into larger cells so that the structure of the finished product remains fine. Thus, two leavening acids with similar leavening rates may produce products with different textures.

2.3 Apparatus and Instruments

1) Filter flask, 125 ml
2) Rubber tubing

3) Rubber stopper
4) Graduated cylinders, 100 and 500 ml
5) Ring stand with holder
6) Magnetic stirrer and stir bars
7) Shallow pan
8) pH meter and pH standard buffers
9) Baking oven set at 218 °C (420 °F)
10) Water bath, 60–70 °C
11) Burettes
12) Electric mixer
13) Thermometer

2.4 Reagents and Materials

1) Sodium bicarbonate
2) Monocalcium phosphate monohydrate (MCP)
3) Sodium aluminum sulfate (SAS)
4) Commercial double-acting baking powder
5) All-purpose flour
6) Sugar
7) Salt
8) Skim milk
9) Vegetable oil
10) Millet seeds

2.5 Procedures

2.5.1 Determination of Leavening Rates

2.5.1.1 The Apparatus
Set up CO_2 measuring apparatus as shown in Figure 2.2.

2.5.1.2 Experimental Treatments and Controls
Note: When there is more than one ingredient, mix them together thoroughly before adding to the flask.

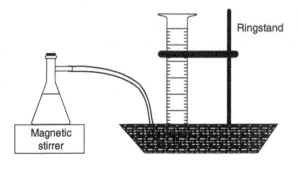

Ringstand

Magnetic stirrer

Figure 2.2 Apparatus for collecting and quantifying the volume of gas released.

1) Control (no leavening agent)
2) Sodium bicarbonate (0.34 g)
3) Sodium bicarbonate (0.34 g) + MCP-H$_2$O (0.42 g)
4) Sodium bicarbonate (0.34 g) + SAS (0.33 g)
5) Baking powder (1.21 g). **Note**: this assumes that baking powder is 28% NaHCO$_3$.

2.5.1.3 Protocol

Note: This is a standard protocol for measuring leavening rates in a model system.

1) Add 60 ml distilled water and a magnetic stir bar to a filter flask.
2) Fill a 100 ml graduated cylinder with tap water and invert it in a pan of water as shown in Figure 2.2 (take care to avoid allowing air to get into the cylinder).
3) Weigh out the specified amounts of each of the leavening agents (see Section 2.5.1.2).
4) With the stirrer running on low speed, transfer the leavening agent to the flask and stopper immediately. (**Note**: The rate of stirring will affect the rate of the reaction, so always use the same setting on the stirrer.)
5) Record the volume of displaced water in the cylinder every 30 seconds for five minutes.
6) Now, set the flask in a water bath held at 65 °C and continue recording the volume for an additional five minutes. (**Note**: Temperature can have a marked effect on CO$_2$ generation, so regulate the temperature carefully.)
7) Measure the final pH of the solution in the filter flask.

2.5.1.4 Data Analysis

Note: A spreadsheet program will save you time on this exercise.

1) For each data point, subtract the control value from the treatment value.
2) Determine the total available CO$_2$ in each of the treatments (express as moles of CO$_2$). Assume that the baking powder contains 28% sodium bicarbonate.
3) Determine the moles of CO$_2$ evolved at each time point. (You will need to use the ideal gas law to calculate the moles of CO$_2$.) Transform your data to % of total available CO$_2$.
4) Plot %CO$_2$ evolved vs time for each of the treatments. Your plot should be similar to Figure 2.1. Be sure to indicate on your plot the point where you put the flask into the 65 °C bath.

2.5.2 Chemically Leavened Biscuits

2.5.2.1 Biscuit Formula

Ingredients	Amount
All-purpose flour	100 g
Sugar	6 g
Sodium bicarbonate	1 g
Leavening acid	To be calculated[a]
Salt	1.5 g
Oil	23 g
Skim milk	80 ml

[a] Use the appropriate neutralizing values to calculate the amount.
Source: Adapted from [11].

2.5.2.2 Treatments
1) Control 1 (no leavening acid and no sodium bicarbonate)
2) Control 2 (1 g sodium bicarbonate but no leavening acid)
3) Fast-acting leavening acid (use monocalcium phosphate monohydrate)
4) Slow-acting leavening acid (use sodium aluminum sulfate)
5) Double-acting leavening acid (use a 50 : 50 mixture of MCP monohydrate and SAS).
6) Baking powder (3.8 g), ***no sodium bicarbonate.***

2.5.2.3 Protocol
1) Mix the dry ingredients thoroughly in a mixing bowl.
2) Add the milk, and oil and mix thoroughly (use 16 strokes) with a rubber spatula.
3) Measure out 65 g aliquots of the dough and place on a baking sheet. Make sure there is 1-inch separation between each biscuit.
4) Bake at 218 °C (420 °F) for 15 minutes.
5) Cool down.
6) Compare volumes (see below) and textures of the various treatments.
7) Calculate the density of each biscuit.

2.5.2.4 Volume Determination of Biscuits
1) Select a plastic container (e.g. margarine tub or yogurt carton) slightly larger than the biscuit.
2) Fill the container level full with fine seeds (rapeseed or millet).
3) Pour seeds in the container into a graduated cylinder and record volume.
4) Place the biscuit in the empty container and refill with seeds.
5) Measure the volume of the seeds and calculate the volume of the biscuit by difference.

2.6 Problem Set

Note: These problems are designed to help you review concepts from introductory chemistry. Consult your introductory chemistry textbook for review.

1 Calculate the volume (at 25 °C and 1 atmosphere) of available CO_2 contained in 0.34 g $NaHCO_3$.

2 Explain why leavening rates differ between leavening acids.

3 The bicarbonate ion contains a proton and yet bicarbonate acts as a base in most leavening systems. Explain why this is so.

4 Why are double-acting baking powders more effective in many baking applications than single-acting baking powders?

5 The recipe for a large batch of pancake batter specifies 1 pound of baking soda and MCP as the leaving acid. How much MCP should be used?

6 Vinegar contains about 5% acetic acid (wt/vol). Calculate the normality of vinegar. What volume of vinegar would be required to neutralize 100 ml of 0.1 N sodium hydroxide?

7 You are baking a batch of biscuits and are out of baking powder. You decide to improvise and use vinegar and baking soda. You decide to add 2 teaspoons of baking soda to your dough. What volume of vinegar will you use? Assume that 1 mole of acetic acid will neutralize 1 mole of baking soda. One teaspoon of baking soda weighs 5 g.

8 The published neutralizing value of monocalcium phosphate monohydrate (MCP) is 80. Calculate the expected NV of monocalcium phosphate monohydrate based on stoichiometric relationships. Assume that the MCP is behaving as a monoprotic acid in this system. Does your result agree with the measured value of 80? If not, give a possible explanation. **Note**: To solve this problem, you will need to write balanced equations for the reactions between MCP and sodium bicarbonate (one with MCP acting as a monoprotic acid and the other with MCP acting as a diprotic acid).

9 What volume of CO_2 would be produced from the complete neutralization of 5 g sodium bicarbonate at 80 °C?

10 Draw the structure of potassium acid tartrate. Write an equation to show why it is a leavening acid.

11 What is the main leavening acid in cultured buttermilk? Draw its structure.

12 What is the main leavening acid in lemon juice? Draw its structure.

13 Explain how glucono-delta-lactone can act as a leavening acid.

14 What is the minimum percentage of sodium bicarbonate in baking powder necessary to yield 12 g of CO_2 per 100 g of powder? (Assume that all of the bicarbonate is converted to CO_2 and H_2O.)

15 What volume of 1.0 molar sulfuric acid would be required to neutralize 100 ml of 1.0 molar sodium hydroxide?

2.7 Useful Formulas and Values

1) Ideal gas law: $PV = nRT$
   where: P = pressure in atmospheres (atm)
   V = volume in liters (l)
   n = number of moles of gas
   $R = 0.0821$ (l) (atm) $(mol^{-1}) (K^{-1})$
   T = temperature in K
2) Volume of 1 mole of CO_2 under standard conditions = 22.40 l.
3) Standard temperature and pressure for gases: $T = 273$ K; $P = 1$ atm.
4) Baking soda content of baking powder: 28% $NaHCO_3$ (w/w).
5) Formula weight of $NaHCO_3 = 84.01$.
6) Molecular weight of $CO_2 = 44.0$.

2.8 References

1 Van Wazer, J.R. and Arvan, P.G. (1954). Chemistry of leavening. *Milling Production* February–March: 3–7.

2 Heidolph, B.B. (1996). Designing chemical leavening systems. *Cereal Foods World* 41 (3): 118–126.

3 Gardner, W.H. (1966). *Food Acidulants*. Allied Chemical. 185 p.

4 Lindsay, R.C. (2017). Food additives. In: *Fennema's Food Chemistry*, 5e (eds. S. Damodaran and K.L. Parkin), 803–864. Boca Raton: CRC Press,Taylor & Francis Group.

5 McGee, H. (2004). *On Food and Cooking: The Science and Lore of the Kitchen*. Completely rev. and updated. New York: Scribner. 884 p.

6 AACC method 2-32 A (1995). Neutralizing value of acid-reacting materials. In: *Approved Methods of the AACC*, 9e. St. Paul, MN: The American Association of Cereal Chemists.

7 Book, S.L. and Waniska, R.D. (2015). Leavening in flour tortillas. In: *Tortillas* (eds. L.W. Rooney and S.O. Serna-Saldivar), 159–183. St. Paul, MN: AACC International Press.

8 Bellido, G.G., Scanlon, M.G., Sapirstein, H.D., and Page, J.H. (2008). Use of a pressuremeter to measure the kinetics of carbon dioxide evolution in chemically leavened wheat flour dough. *Journal of Agricultural and Food Chemistry* 56 (21): 9855–9861.

9 Kichline, T.P. and Conn, T.F. (1970). Some fundamental aspects of leavening agents. *Bakers Digest* 44 (4): 36–40.

10 Conn, J.F. (1981). Chemical leavening systems in flour products. *Cereal Foods World* 26 (3): 119–123.

11 Penfield, M.P. and Campbell, A.M. (1990). *Experimental Food Science*, 3e. San Diego: Academic Press. 541 p. (Food science and technology).

2.9 Suggested Reading

Labaw, G.D. (1982). Chemical leavening agents and their use in bakery products. *Bakers Digest*. 56 (1): 16.

Robinson, J.K., McMurry, J., and Fay, R.C. (2019). *Chemistry*, 8e. Hoboken, NJ: Pearson Education, Inc. 1200 p.

Answers to Problem Set

1 Volume of $CO_2 = 99\,ml$

2 Fast-acting acids dissolve rapidly, slow-acting acids dissolve more slowly.

3 Bicarbonate has a high pK_a, which means it will not dissociate until the pH is high.

4 Double-acting powders allow release of some CO_2 prior to baking which helps improve batter viscosity.

5 1.25 lb MCP.

6 Normality of vinegar = 0.833 N. It will take 12 ml vinegar to neutralize 100 ml of 0.1 N NaOH.

7 143 ml vinegar.

8 Two equations may be written:

Monoprotic:

$$Ca(H_2PO_4)_2H_2O + 2\,NaHCO_3 \longrightarrow CaHPO_4 + Na_2HPO_4 + 2\,CO_2 + 3\,H_2O$$
$$252\,g 168\,g$$

Diprotic:

$$Ca(H_2PO_4)_2H_2O + 4\,NaHCO_3 \longrightarrow CaNaPO_4 + Na_3PO_4 + 4\,CO_2 + 5\,H_2O$$

252 g   336 g

$$NV = b/a \times 100$$

If monoprotic : $NV = 168/252 \times 100 = 67$

If diprotic : $NV = 336/252 = 133$

The published value of 80 lies between 67 and 133. This suggests that the sodium bicarbonate reacts nearly completely with the first H on the $H_2PO_4^-$ but only partially with the second, i.e. the diprotic reaction above does not go to completion.

9 $1.721\,CO_2$.

10

Potassium acid tartrate   Potassium sodium tartrate

11

Lactic acid

12

Citric acid

13   A lactone is an ester formed from the reaction of a carboxyl group and an alcohol group on the same molecule. Glucono-delta-lactone slowly hydrolyzes in water to form gluconic acid.

14   22.9%.

15   50 ml.

3

Properties of Sugars

3.1 Learning Outcomes

After completing this exercise, students will be able to:

1) Draw structures of common reducing and nonreducing sugars.
2) Explain the difference between a hemiacetal and an acetal.
3) Distinguish between reducing and nonreducing sugars experimentally.

3.2 Introduction

Sugars are polyhydroxylated aldehydes or polyhydroxylated ketones (Figure 3.1). Thus, they participate in reactions characteristic of alcohols and aldehydes or ketones. Please review the sections in your organic chemistry textbook that describe reactions for alcohols, aldehydes, and ketones.

Recall that alcohols react reversibly with aldehydes or ketones to form hemiacetals or hemiketals. When the alcohol and carbonyl groups are on the same molecule, as is the case with sugars, a cyclic or ring structure is formed (Figure 3.2).

Note that hemiacetals contain a carbon atom bonded to an –OH group and an –O–R group. Hemiacetals are relatively unstable. In aqueous solution, the open and closed ring forms are both present in equilibrium. Thus, sugars like glucose participate in reactions characteristic of aldehydes even though the predominant form is the hemiacetal.

Sugars containing the hemiacetal group are called reducing sugars because they are capable of reducing various oxidizing agents. Several well-known assays, based on this tendency to oxidize, have been developed for detecting reducing sugars. These include the Tollen's test (sugars are mixed with Ag^+ in aqueous ammonia solution), the Fehling's test (sugars are mixed with Cu^{2+} in aqueous tartrate solution), and the Benedict's test (sugars are mixed with Cu^{2+} in aqueous citrate solution). When mixed with these solutions, reducing sugars are oxidized causing a reduction in the valence of the metal ion. In the Tollen's test, a shiny mirror of elemental silver (Ag^0) forms on the inside surface of the test tube. In the Fehling's and Benedict's tests, Cu^{2+} is reduced to Cu^{1+} which reacts with water to form reddish brown cuprous oxide. Benedict's reagent is used in some of the "sugar sticks" diabetics use to test their urine for spilled sugar. The following chemical equation describes the Benedict's test [1]:

Food Chemistry: A Laboratory Manual, Second Edition. Dennis D. Miller and C. K. Yeung.
© 2022 John Wiley & Sons, Inc. Published 2022 by John Wiley & Sons, Inc.
Companion website: www.wiley.com/go/Miller/foodchemistry2

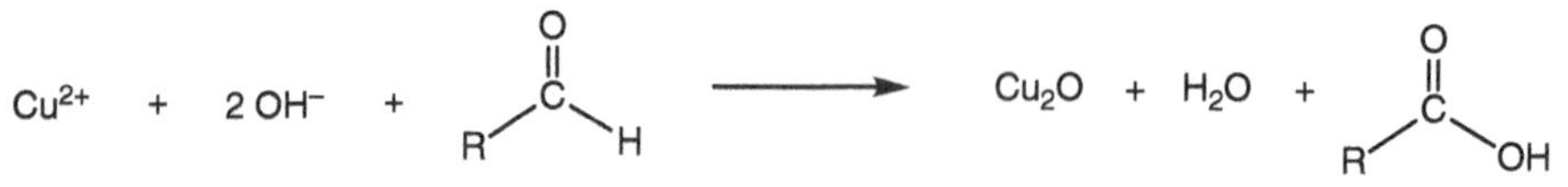

$$Cu^{2+} \quad + \quad 2\ OH^- \quad + \quad \underset{R}{\overset{O}{\underset{\|}{C}}}\!\!-\!\!H \quad \longrightarrow \quad Cu_2O \quad + \quad H_2O \quad + \quad \underset{R}{\overset{O}{\underset{\|}{C}}}\!\!-\!\!OH$$

Cupric ion   Reducing sugar   Cuprous oxide   Aldonic acid

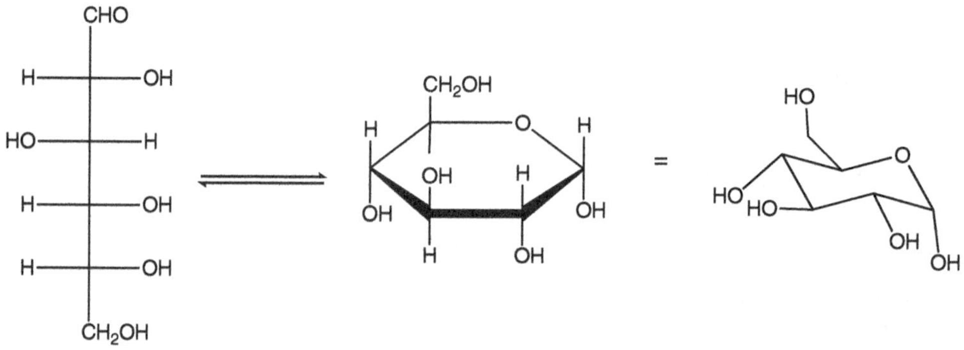

(blue)   (red brown)

D-Glucose   *alpha*-D-Glucose   *alpha*-D-Glucose

(Fisher)   (Haworth)   (Conformational)

Figure 3.1 Three representations of the structure of glucose, a polyhydroxylated aldehyde.

alpha-D-Glucose   *alpha*-D-Glucose

(Open form)   (Ring form)

Figure 3.2 Balanced equation showing the nucleophilic attack of the C-5 hydroxyl oxygen of glucose on the carbonyl carbon of the same molecule to form a hemiacetal.

When conditions are right, hemiacetals can react with alcohols to form acetals. For example, glucose in the hemiacetal form might react with fructose to form the acetal better known as sucrose (Figure 3.3). Carbohydrate acetals are called glycosides.

Recall that acetals contain a carbon bonded to two –O–R groups. Unlike hemiacetals, acetals are relatively stable in aqueous solution and do not readily split into the hemiacetal and alcohol forms. However, they may be split into the alcohol and hemiacetal by acid-catalyzed hydrolysis.

3.3 Apparatus and Instruments

1) Test tubes
2) Volumetric flasks, 50 ml

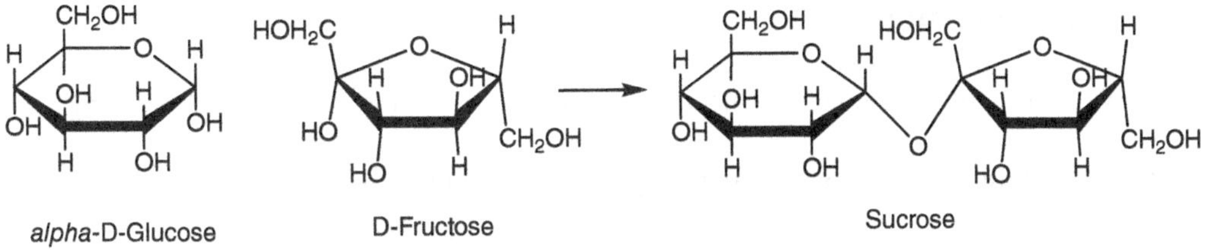

Figure 3.3 The formation of sucrose, a 1,2-glycoside, from glucose and fructose.

3) Hot plate
4) Large beaker, 600 ml
5) Water bath, 37 °C
6) Water bath, boiling

3.4 Reagents and Materials

1) Crystalline glucose, fructose, sucrose, lactose, and sorbitol
2) HCl, 0.5 N
3) NaOH, 1.0 N
   Benedict's solution (CuSO$_4$ in citrate/carbonate buffer) [2].
   To prepare Benedict's solution: Dissolve, with stirring, 17.3 g sodium citrate and 10 g Na$_2$CO$_3$ in 80 ml distilled water. Dissolve 1.73 g CuSO$_4$·5H$_2$O in 10 ml water. Add the CuSO$_4$ solution to the citrate/carbonate solution. Transfer to a 100 ml volumetric flask. Dilute to 100 ml and mix well.
4) Starch solution (3% soluble starch in water)
5) Amylase solution (~1% amylase in water)

3.5 Procedures

1) Transfer 3 ml starch solution to each of two test tubes.
2) Add 5 drops of amylase solution to one of the tubes from Step 1 and incubate both tubes for 15 minutes in a 37 °C water bath.
3) Prepare solutions of glucose, fructose, sucrose, lactose, and sorbitol (50 ml of each, 0.1 mol l^{-1}, in water).
4) Prepare 50 ml of 0.1 mol l^{-1} sucrose in 0.5 N HCl.
5) Transfer 5-ml aliquots of the sucrose-HCl solutions (from Step 4) to two test tubes.
6) Heat one of the test tubes from Step 5 in a boiling water bath for 10 minutes, cool. Keep the other at room temperature. Neutralize the HCl in the tubes by adding 1 ml of 1.0 N NaOH to each tube.
7) Transfer 3 ml of each of the solutions, including the starch solutions and the heated and unheated sucrose solutions, into separate test tubes.
8) Add 8 drops of Benedict's solution to each tube (including the starch solutions and a water blank). Mix well.
9) Place tubes in boiling water bath for three minutes.
10) Remove tubes from bath and observe color and appearance of solutions.

3.6 Study Questions

1 Draw the structure of each of the sugars you tested. Label them reducing or nonreducing.

2 What structure is characteristic of reducing sugars? Explain.

3 Compare your results from the 3 sucrose solutions (in water, in HCl, in HCl with heating). Explain any differences.

4 Compare the structures of glucose and sorbitol (a sugar alcohol).

5 Fructose is not an aldose, yet it is a reducing sugar. Explain. (Hint: Remember that tests for reducing sugars are conducted in alkaline solution and recall the concept of keto-enol tautomerism.)

6 Name and draw the structures of 5 glycosides that may be present in foods.

7 Are all disaccharides nonreducing sugars? Explain.

8 Is starch reducing? Explain. Is starch treated with amylase reducing? Explain.

3.7 References

1 Huber, K.C. and BeMiller, J.N. (2017). Carbohydrates. In: *Fennema's Food Chemistry*, 5e (eds. S. Damodaran and K.L. Parkin), 91–169. Boca Raton: CRC Press,Taylor & Francis Group.
2 Benedict, S.R. (1909). A reagent for the detection of reducing sugars. *Journal of Biological Chemistry* 5: 485–487.

3.8 Suggested Reading

Brady, J.W. (2013). *Introductory Food Chemistry*. Ithaca: Comstock Publishing Associates. 638 p.

4

Nonenzymatic Browning

4.1 Learning Outcomes

After completing this exercise, students will be able to:

1) Prepare buffers and other solutions.
2) Assess and/or predict the extent of browning in a food or beverage based on such factors as type of carbohydrate, protein content, pH, and processing and storage temperatures.
3) Develop formulations of bakery products that optimize nonenzymatic browning.

4.2 Introduction

Brown colors often develop during the processing, storage, and preparation of foods and food ingredients. Some reactions which produce brown colors are enzyme catalyzed. These reactions usually involve oxidation of food components. Other browning reactions are nonenzymatic. These include caramelization of sugars and the Maillard reaction. This laboratory exercise will focus on nonenzymatic browning.

4.2.1 Caramelization

Heating sugars to high temperature causes them to undergo a series of complex decomposition reactions. Products of these reactions include brown-colored polymers and numerous flavor compounds. One of these products, caramel, is widely used to color a variety of beverages and foods including cola, beer, whiskey, bakery goods, and confectionary. In the commercial production of caramels, a sugar, usually sucrose, is mixed with catalytic amounts of food-grade acids, bases, or salts. The choice of catalyst influences the charge, color, and other attributes of the final product [1].

Caramel compounds may be either positively or negatively charged. The charge on the caramel molecules is important because the wrong charge may result in precipitation in the food being colored. For example, caramels used to color soft drinks should be negatively charged so they would not combine with phosphates to cause precipitation. These caramels are produced by heating sucrose in the presence of ammonium bisulfate [2]. The caramel color used in bakery products, beer, and confectionaries are often positively charged [3]. It is produced by adding ammonium salts such as ammonium carbonate or ammonium phosphate [1].

Food Chemistry: A Laboratory Manual, Second Edition. Dennis D. Miller and C. K. Yeung.
© 2022 John Wiley & Sons, Inc. Published 2022 by John Wiley & Sons, Inc.
Companion website: www.wiley.com/go/Miller/foodchemistry2

4.2.2 The Maillard Reaction

The reaction between sugars and amines is known as the Maillard reaction (after the French chemist who studied it). The brown color in Maillard browning is due to the formation of melanoidins which are complex large-molecular-weight molecules. The initial reaction is between an aldehyde or ketone group on a sugar molecule and a free amino group on a protein, peptide, or amino acid molecule, hence the often used term "sugar-amine reaction." The reaction may be desirable (e.g. the chocolate flavor which develops when cocoa beans are roasted is the result of browning) or undesirable (e.g. the objectionable dark brown color that sometimes develops in potato chips during frying). The initial steps in the reaction between a sugar and an amine are shown in Figure 4.1.

The glycosyl amines formed as shown in Figure 4.1 then undergo an Amadori rearrangement to form an amino keto sugar (Figure 4.2).

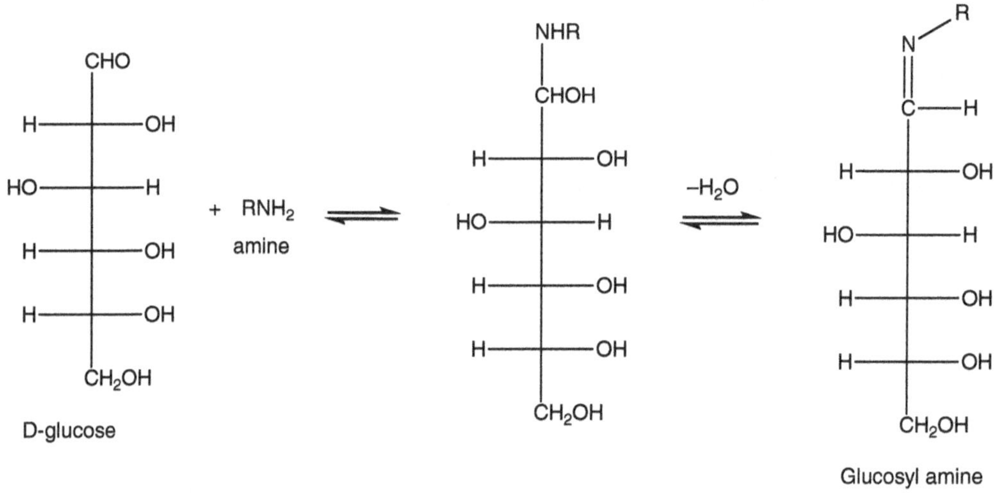

Figure 4.1 Fisher projection formulas showing the reaction of glucose with an amine to form glucosyl amine, an early product of the Maillard reaction. Reducing sugars and proteins or amino acids are common substrates for the Maillard reaction in foods.

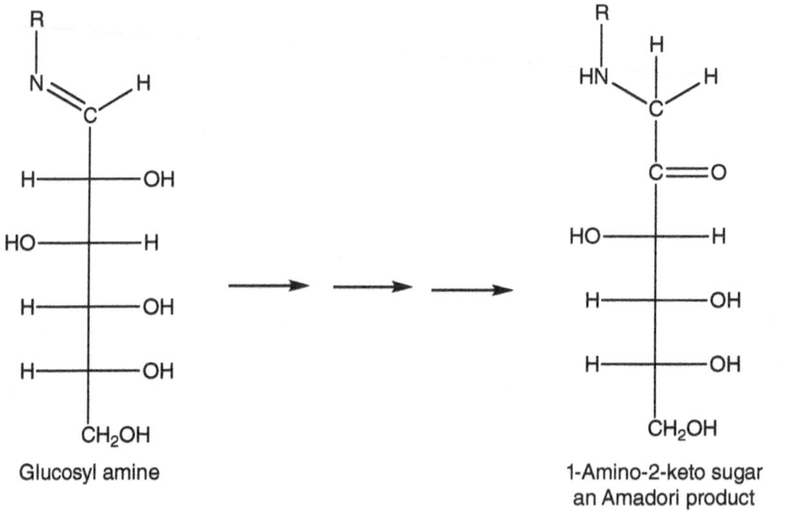

Figure 4.2 The Amadori rearrangement of glucosyl amine to an amino-deoxy-ketose.

Amadori products are unstable to heat and undergo a complex series of reactions that ultimately produce hundreds of flavor and aroma compounds and high molecular-weight brown pigments called melanoidins.

Several factors influence the extent of Maillard browning in a food. First, either an aldehyde or a ketone (reducing sugars are the most important in food systems) and an amine (protein is by far the most important) must be present. Other factors include temperature, concentrations of the sugars and amines, pH, and the type of sugar [1].

4.2.2.1 Sugar

Both stereochemical configuration and size of sugar molecules affect the rate of the Maillard reaction [4]. In general, smaller sugar molecules react faster than larger ones. Pentoses react faster than hexoses and hexoses react faster than disaccharides. Not all hexoses react at the same rate. Galactose seems to be the most reactive among the common hexoses. Fructose reacts more rapidly than glucose in the initial stages but as the reaction proceeds, the rates are reversed [1]. In foods, the sugars most important for Maillard browning are glucose, fructose, maltose, and lactose [5]. In meats, ribose may be significantly involved. See Figure 4.3 for structures of these common sugars.

4.2.2.2 Amine

As mentioned above, proteins, peptides, and free amino acids can participate in the Maillard reaction. N-terminal amino acids in proteins and peptides have a free α-amino group that can react with sugars but the concentration of these is usually low in most foods because proteins are high molecular-weight molecules and there is only one N-terminal amino acid residue per protein molecule. Amino groups tied up in peptide bonds do not react with sugars. Therefore, the ε-amino

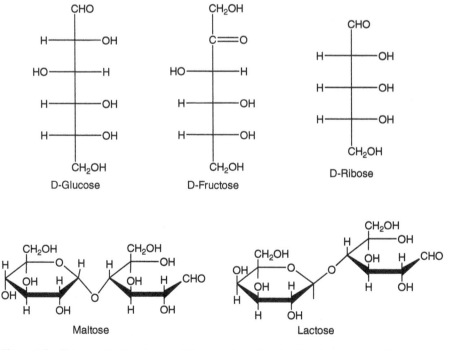

Figure 4.3 Fisher projection (top) and Haworth formulas showing the structure of some common sugars that participate in Maillard browning in foods. Note that they all have free aldehyde or keto groups.

group on lysine residues in proteins and peptides is generally the available amino group in foods present in the highest concentration. See Figure 4.4 for the structures of some amino acid residues in a protein.

4.2.2.3 Temperature
According to Adrian [4], the reaction rate is measurable at 37 °C provided several days of reaction time are allowed, rapid at 100 °C, and violent at 150 °C.

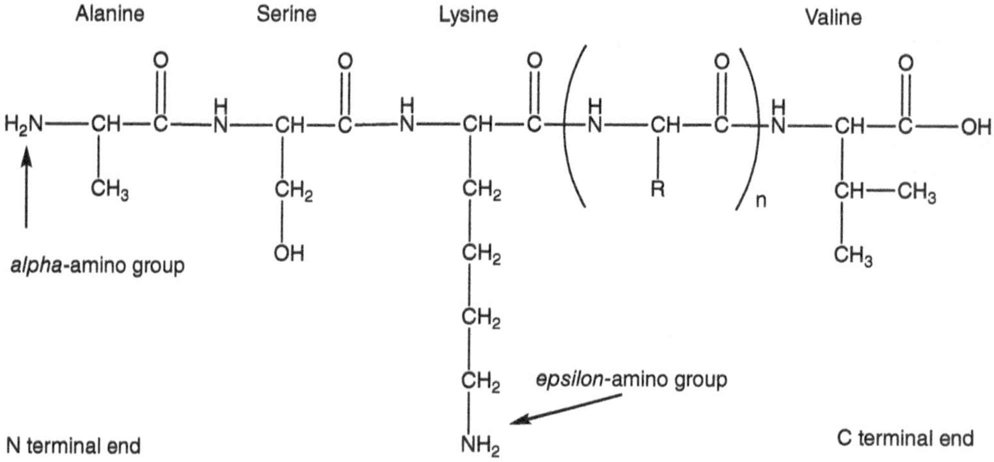

Figure 4.4 Structure of a generic protein showing amino groups that are available for reacting with aldehydes in Maillard browning. Most proteins contain more than 100 amino acid residues, so n will be greater than 100 in most cases. In this example the α-amino group on the amino terminal alanine and the ε-amino group on the lysine residue are available for reaction with sugars.

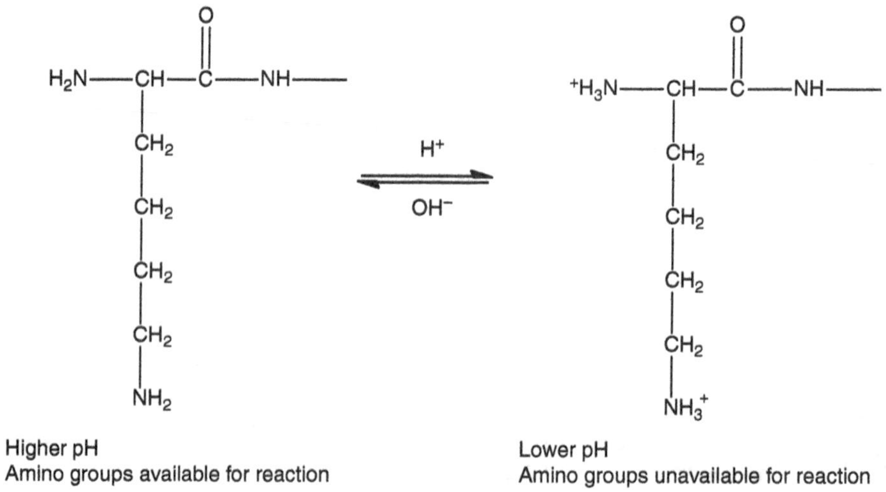

Higher pH
Amino groups available for reaction

Lower pH
Amino groups unavailable for reaction

Figure 4.5 The effect of pH on the protonation of free amino groups in a protein, in this case with an amino terminal lysine residue. In proteins, the pKa for the *terminal* α-amino group is around 7.8 while it is approximately 10.2 for the ε-amino group on a lysine residue.

Table 4.1 Effects of storage on skim milk powder[a,b].

	Fresh powder	Stored powder
pH of reconstituted milk	6.73	6.50
Reducing power (ferricyanide value)	0.9	16.0
Free amino-N content (% initial value)	100	36
Biological value (protein)	84.5	67.5
Biological value with added lysine	76.4	80.1
Flavor of reconstituted milk	Palatable	Nauseating

[a] Storage conditions: 60 days, 37 °C, 7.3% moisture.
[b] Modified from [4].

4.2.2.4 Concentration

The reaction is extremely slow in very dry foods and in highly dilute solutions [4]. Maximum rates of browning reactions occur in foods containing about 10–15% moisture. This is because some water is necessary for the reactants to interact but in very dilute solutions, the reactants would be relatively widely separated. Water may also act as an inhibitor of the reaction since several of the steps in the complex series of reactions are dehydrations and an excess of water, a product of dehydration reactions, could be expected to inhibit the reaction.

4.2.2.5 pH

The major effect of pH is related to the protonation of amino groups. At low pH, more of the amine groups would be protonated and fewer would be available for reaction (Figure 4.5).

Maillard browning is a common problem in stored nonfat dried milk because of its high lactose content and its reactive protein. Storage of milk powders under unfavorable conditions can result in serious deterioration in quality. Table 4.1 shows a comparison of fresh and stored milk powder.

4.3 Apparatus and Instruments

1) pH meter
2) Pipettes, 10 ml
3) Volumetric flasks, 50 and 100 ml
4) Beakers, 50, 100, 250, and 600 ml
5) Glass stirring rods
6) Test tubes
7) Test tube racks
8) Permanent marker
9) Hot plate
10) Graduated cylinders, 10 and 50 ml
11) Pasteur pipettes
12) Vortex mixer
13) Top loading balance
14) Spectrophotometer
15) Water bath, 95 °C

16) Aluminum weighing dishes
17) Household bread machine
18) Oven, set at 125 °C for nonfat dry milk, and at 190 °C for cookies
19) Cookie sheets
20) Mixing bowls and utensils for baking

4.4 Reagents and Materials

1) Crystalline glucose
2) Crystalline glycine
3) KH_2PO_4, 1/15 M
4) Na_2HPO_4, 1/15 M
5) Nonfat dry milk
6) Ingredients for yeast bread (flour, water, sugar, salt, and yeast)
7) Ingredients for cookies (flour, shortening, baking soda, baking powder, egg, sucrose)
8) High fructose corn syrup
9) Corn syrup

4.4.1 Reagents to Be Prepared by the Student

1) Phosphate buffer, 1/15 M, pH 5 and 8 (see Appendix III for buffer tables)
2) Glucose, 0.5 M, in phosphate buffer, pH 5 and 8
3) Glycine, 0.5 M, in phosphate buffer, pH 5 and 8

4.4.2 Reagents to Be Prepared by the Teaching Staff

1) Fructose, 0.25 M + 0.25 M glycine in 1/15 M phosphate buffer, pH 5 and 8
2) Sucrose, 0.25 M + 0.25 M glycine in 1/15 M phosphate buffer, pH 5 and 8
3) Lactose, 0.25 M + 0.25 M glycine in 1/15 M phosphate buffer, pH 5 and 8
4) Glucose, 0.25 M in 1/15 M phosphate buffer, pH 5 and 8
5) Fructose, 0.25 M in 1/15 M phosphate buffer, pH 5 and 8
6) Sucrose, 0.25 M in 1/15 M phosphate buffer, pH 5 and 8
7) Lactose, 0.25 M in 1/15 M phosphate buffer, pH 5 and 8
8) Glycine, 0.25 M in 1/15 M phosphate buffer, pH 5 and 8

4.5 Procedures

4.5.1 Preparation of a Glucose/Glycine Model System

Note: The objective here is to prepare buffered glucose/glycine solutions with identical concentrations but varying pHs.

1) Prepare 100 ml of 1/15 M phosphate buffer, pH 5.0 and 100 ml of 1/15 M phosphate buffer, pH 8.0 (prepare from 1/15 M KH_2PO_4 and 1/15 M Na_2HPO_4). See Appendix for volumes to mix.
2) Prepare 50 ml of 0.5 M glucose solutions in each of the phosphate buffers. (The M.W. of the glucose is 180.16 g $mole^{-1}$.) Add the glucose a little at a time with stirring to about 30 ml of the buffer in a beaker, transfer to a volumetric flask, and dilute to volume with buffer.

3) Prepare 50 ml of 0.5 M glycine solution in each of the phosphate buffers. (The M.W. of glycine is 75 g mole^{-1}.)
4) In a test tube, mix 5 ml of the glucose solution with 5 ml of the glycine solution to form a glucose-glycine solution. What are the molar concentrations of glucose and glycine in this solution? Do this for both pH 5 and 8.
5) Heat the solutions in a 95 °C water bath for 30 minutes (see Section 4.5.2).

4.5.2 Heating Experiment

1) Label 18 test tubes with the treatments shown below and place them in plastic test tube racks. Transfer 10 ml aliquots of the solutions listed below to the test tubes and cap them loosely.

Treatment at pH 5	Treatment at pH 8
Glucose	Glucose
Fructose	Fructose
Sucrose	Sucrose
Lactose	Lactose
Glycine	Glycine
Glucose/glycine	Glucose/glycine
Fructose/glycine	Fructose/glycine
Sucrose/glycine	Sucrose/glycine
Lactose/glycine	Lactose/glycine

2) Place all tubes in a 95 °C water bath for 30 minutes.

4.5.3 Measurement of Extent of Browning

1) After the tubes from the 95 °C water bath have cooled, measure the pH in each tube.
2) Turn on your spectrophotometer and allow it to warm up. Turn the wavelength selector to 430 nm. Use water to set 0 absorbance.
3) Measure the absorbance of each of your solutions. You may have to dilute (with water) the darker solutions to keep them on scale. To calculate the absorbance of the original undiluted solutions, multiply the absorbance of the diluted solution by the dilution factor.

4.5.4 Browning in Nonfat Dry Milk (Demonstration)

1) Cover the bottom of 5 aluminum weighing dishes with nonfat dry milk.
2) Place 4 of the samples in a 125 °C oven.
3) Remove one sample at 10, 20, 30, and 60 minutes.
4) Compare all samples and describe the color of the samples.

4.5.5 Role of Milk in Crust Color of Bread (Demonstration)

Many recipes for bread call for the addition of nonfat dry milk to promote Maillard browning in the crust. In this experiment, we will test the hypothesis that the crust color of bread is darker when bread is made with a small amount of nonfat dry milk compared with bread made without milk.

The basic formula for yeast bread is flour, water, sugar, salt, and yeast. Small amounts of fat and milk may also be added. Most commercial breads contain anti-staling agents such as mono or di-glycerides and mold-inhibiting preservatives such as sodium or potassium propionate.

A household bread machine will be used to bake the bread. Add the ingredients to the bread pan in the amounts and order specified in the manual provided by the manufacturer of the bread machine. Start the machine. Do not open the lid on the machine until the cycle is complete. When the baking cycle is complete, remove the pan from the machine (use oven mitts) and remove the loaf from the pan.

Compare the crust color of loaves prepared with and without the added nonfat milk powder. Slice the bread and compare the crumb structure of the breads.

4.5.6 Browning in Cookies

Most cookies contain high concentrations of sugar, usually sucrose. The type and concentration of sugar in cookies affects cookie spread during baking, surface cracking, and browning, among other quality factors. In this experiment, we will make sugar cookies using a standard recipe and then substitute either high-fructose corn syrup or regular corn syrup for a portion of the sucrose. (In the cookie formula below, reduce the sucrose to 65 g and add 10 g of either high-fructose corn syrup or regular corn syrup.)

4.5.6.1 Sugar Cookie Formula

Ingredients	Amount
Flour (g)	85
Shortening (g)	55
Sucrose (g)	75
Baking soda (g)	1
Baking powder (g)	0.6
Egg	¼ egg

Source: From http://allrecipes.com/recipe/easy-sugar-cookies/?scale=12&ismetric=0.

4.5.6.2 Baking Directions

1) Preheat oven to 375 °F (190 °C). In a small bowl, stir together flour, baking soda, and baking powder. Set aside.
2) In a large bowl, cream together the butter and sugar until smooth. Beat in egg and vanilla. Gradually blend in the dry ingredients. Roll rounded teaspoonfuls of dough into balls, and place onto an ungreased cookie sheet.
3) Bake 8–10 minutes in the preheated oven, or until golden. Let stand on the cookie sheet for two minutes before removing to cool on wire racks.
   Prepare three batches:
   - Control: As above with 75 g sucrose
   - High-fructose corn syrup: 65 g sucrose, 10 g HFCS
   - Corn syrup: 65 g sucrose, 10 g corn syrup
   Source: From http://allrecipes.com/recipe/easy-sugar-cookies/?scale=12&ismetric=0.

4.6 Problem Set

1 Describe carefully how you would prepare 1 l of 1/15 M phosphate buffer, pH 8 from crystalline, anhydrous KH_2PO_4 and Na_2HPO_4. You may use the tables for buffer preparation in Appendix III.

2 Describe how you would prepare 100 ml of a solution containing 0.25 M glucose and 0.25 M glycine in water.

3 Nonfat fluid milk contains approximately 2.7 g of lysine (2.37 g lysine residues) per liter, almost all of which is present as amino acid residues in casein and whey protein. (**Note**: the molecular weight of lysine is 146 but the molecular weight of a lysine residue is 128. Can you explain the difference?) Assuming that the pH of milk is 6.6, calculate the concentration of un-protonated ε-amino groups in milk. Recall that the pK_a for lysine residues in a protein is approximately 10.2. **Hint**: use the Henderson–Hasselbalch equation. Based on your answer, would you expect lysine residues in milk protein to participate in Maillard browning? Why or why not?

4.7 Study Questions

1 Explain the difference in browning that you observed between the glucose/glycine and the sucrose/glycine model systems.

2 Explain the effect of pH on Maillard browning.

3 If you were to manufacture a formulated food product containing added sugar, what (a) ingredients, (b) processing techniques, and (c) storage conditions would you use to minimize nonenzymatic browning? How might you enhance nonenzymatic browning?

4 Give three examples each of desirable and undesirable nonenzymatic browning reactions in food systems.

4.8 References

1 Huber, K.C. and BeMiller, J.N. (2017). Carbohydrates. In: *Fennema's Food Chemistry*, 5e (eds. S. Damodaran and K.L. Parkin), 91–169. Boca Raton: CRC Press, Taylor & Francis Group.
2 Myers, D.V. and Howell, J.C. (1992). Characterization and specifications of caramel colours: an overview. *Food and Chemical Toxicology* 30 (5): 359–363.
3 Wong, D. (2017). *Mechanism and Theory in Food Chemistry*, 2e. New York, NY: Springer Science+Business Media. 450 p.
4 Adrian, J. (1982). The Maillard reaction. In: *Handbook of Nutritive Value of Processed Food Volume 1 Food for Human Use* (ed. M. Rechcígl), 592–608. Boca Raton, FL: CRC Press.
5 Sikorski, Z.E., Pokorny, J., and Damodaran, S. (2008). Physical and chemical interactions of components in food systems. In: *Fennema's Food Chemistry*, 4e (eds. S. Damodaran, K. Parkin and O.R. Fennema), 849–883. Boca Raton: CRC Press/Taylor & Francis.

4.9   Suggested Reading

Belitz, H.-D., Grosch, W., and Schieberle, P. (2009). *Food Chemistry*, 4e. Berlin: Springer. 1070 p.

BeMiller, J.N. and Whistler, R.L. (eds.) (2009). *Starch: Chemistry and Technology*, 3e. London: Academic Press. 879 p. (Food science and technology).

Brady, J.W. (2013). *Introductory Food Chemistry*. Ithaca: Comstock Publishing Associates. 638 p.

Buffer Reference Center (2020). Sigma-Aldrich [Internet]. https://www.sigmaaldrich.com/life-science/core-bioreagents/biological-buffers/learning-center/buffer-reference-center.html (accessed 10 February 2020.

Reyes, F.G.R., Poocharoen, B., and Wrolstad, R.E. (1982). Maillard browning reaction of sugar-glycine model systems: changes in sugar concentration, color and appearance. *Journal of Food Science* 47 (4): 1376–1377.

Sapers, G.M. (1993). Browning of foods: control by sulfites, antioxidants, and other means. *Food Technology* 47 (10): 75–84.

Shallenberger, R.S. and Birch, G.G. (1975). *Sugar Chemistry*. Westport, CT: Avi Pub. Co. 221 p.

Answers to Problem Set

1 Dissolve 3.74 g KH_2PO_4 and 5.56 g Na_2HPO_4 and bring the volume to 1 l in a volumetric flask.
2 Dissolve 4.5 g glucose and 1.89 g glycine and bring the volume to 100 ml in a volumetric flask.
3 Concentration of un-protonated ε-amino groups in milk = 0.005 mmol L^{-1}.

5

Food Hydrocolloids

5.1 Learning Outcomes

After completing this exercise, students will be able to:

1) Describe the chemistry related to the functional proprieties of selected food hydrocolloids.
2) Prepare dispersions of food hydrocolloids in aqueous systems.
3) Measure the viscosity of food hydrocolloid dispersions.
4) Compare the properties of different food gums under conditions that may be present in foods.
5) Select an appropriate hydrocolloid for a specific food application.

5.2 Introduction

Hydrocolloids are polymers that can be dissolved or dispersed in water and that produce thickening or gelling. Most food hydrocolloids are polysaccharides although some proteins (e.g. gelatin) also fit the definition. *Hydrocolloid* is the scientifically preferred term for these materials but *gum* is a common synonym and *mucilage* is also used. Hydrocolloids are used extensively as food additives to perform a variety of functions (Table 5.1).

The basis for many of the functional properties of hydrocolloids is their remarkable capacity to increase viscosity (thicken) and form gels in aqueous systems at low concentrations. Polysaccharide hydrocolloids vary in molecular weight, chain branching, charge, and hydrogen bond-forming groups. The effectiveness of hydrocolloids in providing functionality to foods varies with the hydrocolloid and the food. Thus, food technologists must be able to select the right hydrocolloid for a specific application.

Polysaccharide hydrocolloids may be linear or branched. In general, thickening power increases with molecular weight and decreases with chain branching. This is because extended straight-chain molecules "sweep" out a larger volume as they tumble in solution than branched-chain molecules which are more compact. Thus, large linear polymers with their associated water collide more frequently with each other than branched and lower molecular-weight molecules (Figure 5.1). These collisions create friction which impedes the flow of the solution and causes an increase in viscosity.

Most hydrocolloids tend to form clumps when the powdered hydrocolloid is mixed with water. Since hydrocolloids must be in solution in order to provide desired functionalities, it is imperative

Food Chemistry: A Laboratory Manual, Second Edition. Dennis D. Miller and C. K. Yeung.
© 2022 John Wiley & Sons, Inc. Published 2022 by John Wiley & Sons, Inc.
Companion website: www.wiley.com/go/Miller/foodchemistry2

Table 5.1 Some functional properties of food hydrocolloids[a].

Function	Examples of food applications
Fat replacer	Low-fat ice creams, fat-free cream cheese, reduced-fat salad dressings
Adding body	Low-calorie beverages
Inhibition of sugar crystallization	Ice cream, syrups
Clarification	Beer, wine
Clouding	Fruit drinks
Dietary fiber	Breakfast cereals
Emulsification	Salad dressings, sauces
Gelling	Puddings
Stabilization	Salad dressings, ice cream
Particle suspension	Chocolate milk
Thickening	Jams, pie fillings, sauces

[a] Modified from [1].

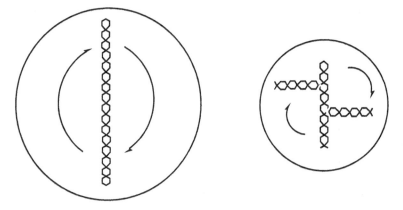

Figure 5.1 Two polymers of equal molecular weight tumbling in solution. Note how the branched-chain polymer occupies less space in solution.

that mixing with water is done properly. Gradual addition of powders to water with high-shear mixing is one way to avoid clumping. Another is to mix the dry hydrocolloid with a liquid "non-solvent" such as vegetable oil, alcohol, or corn syrup before mixing with water [2].

Alginate, carrageenan, locust bean gum (LBG), guar gum, and xanthan gum are hydrocolloids that are widely used in foods. A brief description of their structures and properties follows. See references by Glicksman [1, 3], King [2], and Pettitt [4] for an in-depth discussion of the chemistry, properties, and food applications of hydrocolloids.

5.2.1 Alginate

Alginates are salts of alginic acid. They are polysaccharides composed primarily of mannuronic and guluronic acids joined by 1 → 4 linkages. Alginates occur naturally in brown seaweeds where they are present as insoluble mixed salts of calcium, magnesium, sodium, and potassium [2].

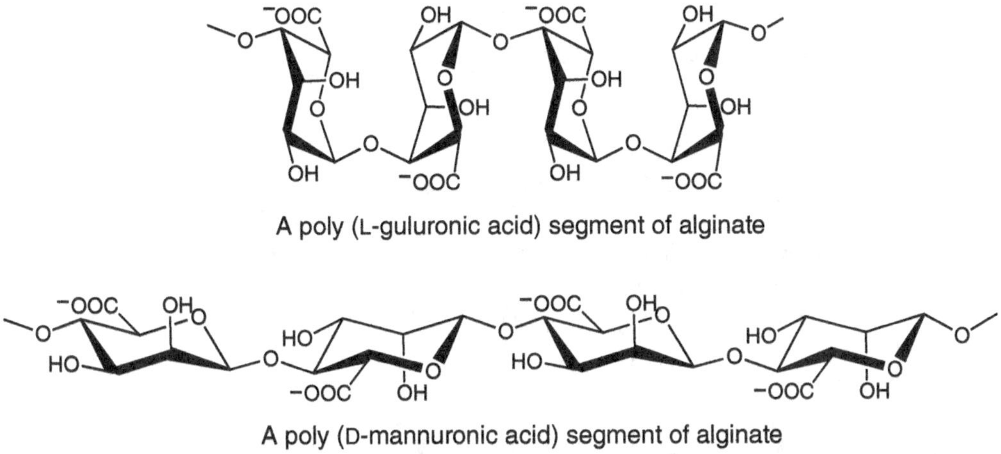

A poly (L-guluronic acid) segment of alginate

A poly (D-mannuronic acid) segment of alginate

Figure 5.2 Conformation of linkages between guluronate and mannuronate residues in alginate.

Sodium and potassium salts of alginic acid are water soluble. Thus, alginates may be extracted from seaweeds with dilute alkali.

The uronic acid residues are arranged in blocks of mannuronic acid, guluronic acid, and alternating mannuronic/guluronic residues: -M-M-M-M-, -G-G-G-G-, and -M-G-M-G-M-G-. The polymers are unbranched. Mannuronic acid residues are linked β-1,4 while guluronic residues are linked α-1,4 (Figure 5.2).

Alginates are available commercially in the form of ammonium, calcium, sodium or potassium salts of alginic acid, or as propylene glycol alginate. The viscosity of alginate solutions is affected by the type of cation present in the solution. Viscosity is relatively low when monovalent cations like Na^+ and K^+ are present but increases in the presence of di- and trivalent cations. Most noticeably, alginate solutions form an irreversible gel with calcium in cold water, provided calcium availability is controlled to avoid precipitation. Alginates are noted for their gelling and thickening properties as well as their ability to stabilize emulsions and inhibit syneresis. Sodium alginate, when added to a wheat starch suspension at 2% w/w, has also been shown to significantly retard retrogradation making it capable of potentially extending the shelf life of wheat-based foods [5]. Alginates are used in salad dressings, pie fillings, structured fruit pieces, and many bakery and dairy products.

5.2.2 Alginate Gels

Alginates are very versatile gelling agents and may be used to form gels in many different kinds of foods. Alginates will form gels over a wide range of pH values and gelling occurs readily without heating. Calcium is required for gel formation. Control of the free calcium concentration is key to successful gelation. If soluble calcium salts such as $CaCl_2$ are added too rapidly to a solution of alginate, precipitates rather than gels will form. A proposed model for explaining how Ca^{2+} produces gels with alginate is the so-called egg box structure shown in Figure 5.3.

Two approaches to forming alginate gels are known as "diffusion setting" and "internal setting" [2]. Diffusion setting involves mixing solutions of calcium and alginate. When the calcium from the calcium solution contacts the alginate, gel formation is very rapid at the interface of the two solutions. Gel strength increases with time as more calcium diffuses into the interior of the gel.

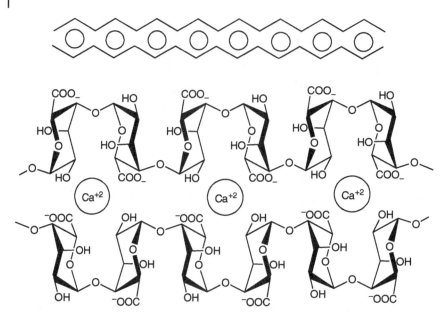

Figure 5.3 The unique conformation of the guluronate linkages in alginate allows for efficient cross linking with polyvalent cations such as Ca^{2+}. The resulting complexes are commonly called "egg box" structures since schematic representations resemble an egg box as shown in the upper part of the figure [6].

The diffusion setting method is used to make extruded foods such as onion rings, pimento strips for olives, and artificial fruit pieces. It is also used to coat foods with alginate films to prevent freezer burn and moisture loss. Internal setting, as the name implies, involves the controlled release of calcium from calcium complexes dispersed in the alginate solution. Gels formed by internal setting are more uniform than gels prepared by the diffusion setting method. They may be formed by thoroughly mixing all dry ingredients including the alginate and a calcium source and then adding the mixture to water. Calcium release into solution is controlled either by using a slowly soluble calcium salt or by using calcium sequestrants. The rate of calcium release may be manipulated by raising the temperature or lowering the pH. Rates of solubilization increase with increasing temperature while lowering the pH causes dissociation of calcium from complexes, thereby freeing it to interact with the alginate.

5.2.3 Carrageenan

Carrageenans are a family of sulfated linear polysaccharides that occur naturally in marine red algae, a common seaweed in the North Atlantic. They are polymers of D-galactose and 3,6-anhydro-D-galactose. Members of the family differ in the degree and location of the sulfate ester groups and the relative proportions of D-galactose and 3,6-anhydro-D-galactose. Sulfate content varies from 18 to 40% [3]. The three principal fractions of carrageenan are kappa (gelling), iota (gelling), and lambda (non-gelling) (Figure 5.4). Carrageenan gels are thermo-reversible, an important property in some food applications.

Carrageenans are widely used as food additives in dairy products and sauces to improve texture and viscosity. In addition to food applications, durable superabsorbent hydrogels prepared from

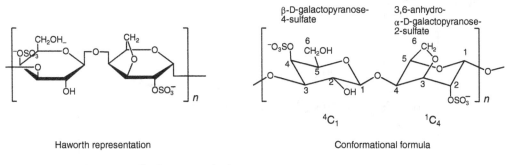

Basic repeat unit of κ-carrageenan

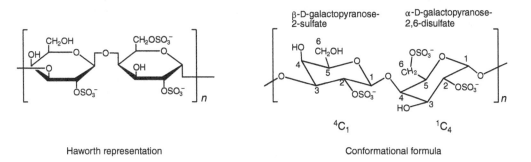

Basic repeat unit of ι-carrageenan

Basic repeat unit of λ-carrageenan

Figure 5.4 Structures of the three principal carrageenans [6]. Note differences in the extent and position of the sulfation.

aqueous solutions containing 4–5% kappa-carrageenan [7] could potentially be used as soil conditioners and carriers for nutrient release [8].

5.2.4 Locust Bean Gum and Guar Gum

LBG is a water extract of the endosperm of seeds (locust beans) of the carob tree. Thus, LBG is also known as carob bean gum. These leguminous trees have been cultivated for centuries along the coast of the Mediterranean Sea. It is believed that one of the first uses of LBG was to bind together the cloths used to wrap mummies in ancient Egypt. LBG is a galactomannan. It consists of a long backbone chain composed of repeating mannose units linked by β-1,4-glycosidic bonds with galactose side chains linked to mannose residues by an α-1,6-bond (Figure 5.5) [9]. Another common galactomannan is guar gum, which is extracted from guar beans and has a lower mannose to galactose ratio compared with LBG.

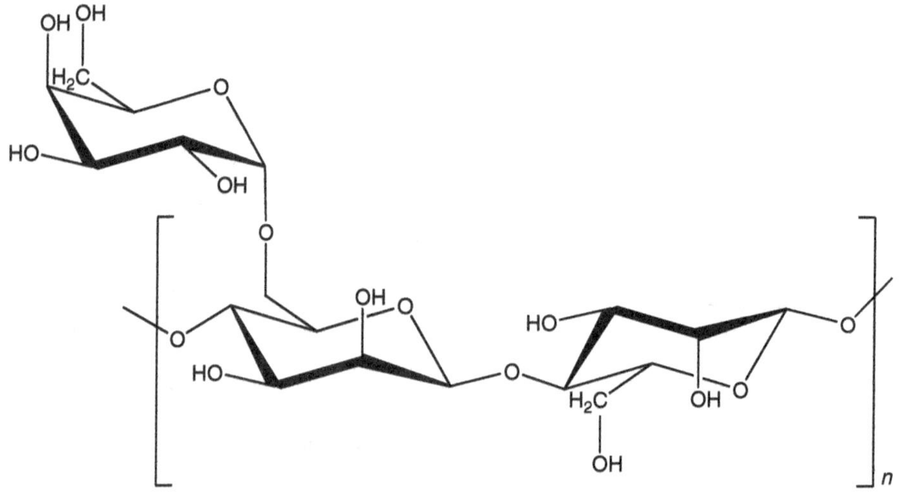

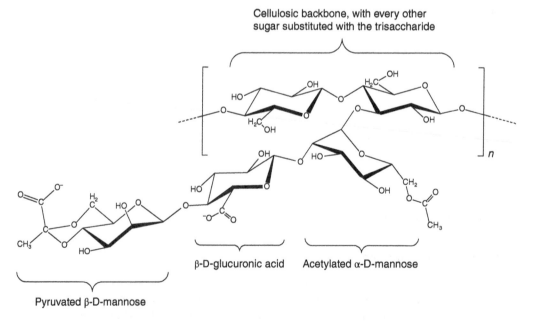

Figure 5.5 A representative structure of galactomannan. The backbone chain is composed of repeating mannose residues linked by β-1,4-glycosidic bonds. The side chains are single galactose units linked to the backbone by α-1,6-bonds.

Cellulosic backbone, with every other sugar substituted with the trisaccharide

β-D-glucuronic acid   Acetylated α-D-mannose

Pyruvated β-D-mannose

Figure 5.6 Structure of xanthan gum. Note the cellulose backbone and the trisaccharide side chain. The sugars in the side chain are mannose-glucuronic acid-mannose. The first mannose is esterified to acetic acid on C-6. The end mannose is linked to pyruvic acid by a ketal linkage at C-4 and C-6 [6].

Table 5.2 Functional properties of xanthan gum [4].

- High solubility in cold and hot water
- Soluble and stable over a wide pH range
- Stable to heat
- Imparts high viscosity to aqueous systems at low concentrations
- Undergoes shear thinning when a shear stress is applied, e.g. when a salad dressing is poured
- Viscosities are uniform over the temperature range of 0–100 °C
- Compatible with most salts present in foods, e.g. does not gel in the presence of Ca^{2+}

Galactomannans are often included in food formulations as stabilizers. For example, LBG has been shown to reduce the melting rate of ice-cream [10] and improve the hardness and adhesiveness of spreadable ricotta cheese [11].

5.2.5 Xanthan Gum

Xanthan gum is an extracellular polysaccharide produced by the bacterium *Xanthomonas campestris*. *X. campestris* was originally isolated from rutabaga plants. It is now grown in pure culture for the express purpose of producing xanthan gums for food and industrial uses. Xanthan gum has a straight-chain backbone identical to cellulose (β-1,4 linked glucose residues). It differs from cellulose in that it contains trisaccharide branches on the number 3 carbon of alternating glucose residues in the backbone chain (Figure 5.6).

The negatively charged branches on the backbone chain lend rigidity to the polymers which have a very high molecular weight estimated to be about 15 million [4]. This structure explains some of the unique properties of xanthan gums, summarized in Table 5.2. These properties make xanthan gums very attractive for sauces and salad dressings because the high viscosity prevents sedimentation of particulates and creaming of oils while the shear thinning ensures pourability.

5.3 Apparatus and Instruments

1) Brookfield viscometer (LVF) (spindle no. 3, 60 rpm)
2) Hot plate with magnetic stirrer and stir bars
3) Balances and weighing paper
4) Beakers, 250 and 600 ml
5) Graduated cylinders, 100 and 250 ml
6) Household mixer
7) Stirring rod
8) Watch glass
9) Screw-top test tubes and caps
10) Thermometer

5.4 Reagents and Materials

1) Hydrocolloids: sodium alginate, LBG, and xanthan gum. **Note**: TIC Gums Pretested colloid 488 T powder (sodium alginate) works well.
2) Glycerol

3) Sodium hexametaphosphate (powder)
4) Calcium hydrogen phosphate, anhydrous (CaHPO$_4$)
5) CaCl$_2$·2H$_2$O
6) Food coloring

5.5 Procedures

Note: Do not discard the gum solutions until the entire exercise is completed.

5.5.1 Effect of Heat Treatment on Gelation

1) Prepare 2 solutions of pre-hydrated LBG and pre-hydrated xanthan gum as follows:
    a) Transfer 100 ml distilled water to each of two 250 ml beakers. Begin stirring gently on a stir plate and, while stirring, add 0.25 g of LBG to each beaker. Repeat the step with 0.25 g xanthan gum.
    b) Cover one of the beakers with a watch glass and heat to boiling on a hot plate while stirring gently.
    c) Place both solutions in the refrigerator.
2) After two hours, observe the consistency of both solutions.

5.5.2 Effect of Concentration on Viscosity

1) Prepare 800 ml of 10 mmol l^{-1} CaCl$_2$·2H$_2$O in deionized water and transfer to a mixing bowl.
2) Weigh out 6.0 g of sodium alginate powder.
3) Mix the sodium alginate with 30 ml glycerol.
4) Turn on the mixer at low to medium speed.
5) With the mixer running, pour the gum/glycerol mixture *slowly* into the CaCl$_2$ solution and continue mixing until all of the gum is hydrated.
6) Transfer 400, 266, and 133 ml of the alginate solution to separate 600 ml beakers. Bring the total volume in each beaker to 400 ml with distilled water. (What are the resulting gum concentrations?)
7) Measure the viscosity of each solution with a Brookfield viscometer using a no. 3 spindle and record the results.
8) Add 4 g of sodium hexametaphosphate to each of the sodium-calcium alginate solutions, mix, and repeat the viscosity measurement. (Sodium hexametaphosphate is a calcium sequestering agent.)
9) Repeat Steps 1 through 6 using xanthan gum.
10) Make plots of viscosity (cP) versus concentration (% gum, w/v).

5.5.3 Emulsion Stability

1) Mark 3 test tubes at the 5 ml level.
2) Pour 5 ml of each of the following into separate test tubes: water, 0.5% sodium alginate, and 0.5% xanthan.
3) Add 5 ml of vegetable oil to each test tube.
4) Shake each test tube vigorously for 30 seconds.
5) Record time required for the water phase and the oil phases to separate.

5.5.4 Diffusion Setting and Internal Setting Alginate Gels

This procedure was adapted from King [2].

5.5.4.1 Diffusion Setting Gel

1) Transfer 236 ml cold water to a 400 ml beaker. Thoroughly mix 45 g fine granulated sugar and 1.7 g low-residual-calcium sodium alginate. Gradually add the dry ingredients, stirring constantly, to the water.
2) When the dry ingredients are completely dissolved, add 5 drops of food coloring and 1.0 ml 2.6 M $CaCl_2$ (0.1 g calcium) to the solution. Stir for one minute. Cover the beaker and leave at room temperature overnight.

5.5.4.2 Internal Setting Gel

1) Transfer 236 ml cold water to a 400 ml beaker and add 5 drops of food coloring. Combine 45 g fine granulated sugar, 1.7 g low-residual-calcium sodium alginate, 1.6 g food-grade adipic acid, 1.9 g sodium citrate, and 0.18 g anhydrous calcium hydrogen phosphate ($CaHPO_4$) (0.18 g $CaHPO_4$ contains 0.1 g calcium). Mix thoroughly.
2) Add the dry ingredients to the water and stir briskly for one minute. Cover the beaker and leave at room temperature overnight.

Compare the appearance, strength, and texture of the two gels.

5.6 Study Questions

1 Define the term viscosity and explain why it is an important property of foods.

2 Describe the operation of a Brookfield viscometer and explain how readings from the instrument are converted to units of viscosity (cps).

3 Xanthan gum is not considered an emulsifying agent, yet it is very effective at stabilizing salad dressings. Explain.

4 Alginate is a calcium-sensitive hydrocolloid. What does this mean and why, given its structure, might you expect it to be calcium sensitive?

5.7 References

1 Glicksman, M. (1982). Origins and classification of hydrocolloids. In: *Food Hydrocolloids* (ed. M. Glicksman), 4–18. Boca Raton, FL: CRC Press.
2 King, A.H. (1982). Brown seaweed extracts (alginates). In: *Food Hydrocolloids* (ed. M. Glicksman), 116–181. Boca Raton, FL: CRC Press.
3 Glicksman, M. (1982). Red seaweed extracts (agar, carrageenans, furcellaran). In: *Food Hydrocolloids* (ed. M. Glicksman), 83–107. Boca Raton, FL: CRC Press.
4 Pettitt, D.J. (1982). Xanthan gum. In: *Food Hydrocolloids* (ed. M. Glicksman), 128–146. Boca Raton, FL: CRC Press.

5 Yu, Z., Wang, Y.-S., Chen, H.-H. et al. (2018). The gelatinization and retrogradation properties of wheat starch with the addition of stearic acid and sodium alginate. *Food Hydrocolloids* 81: 77–86.

6 Brady, J.W. (2013). *Introductory Food Chemistry*. Ithaca: Comstock Publishing Associates. 638 p.

7 Berton, S.B.R., de Jesus, G.A.M., Sabino, R.M. et al. (2020). Properties of a commercial κ-carrageenan food ingredient and its durable superabsorbent hydrogels. *Carbohydrate Research* 487: 107883.

8 Guilherme, M.R., Aouada, F.A., Fajardo, A.R. et al. (2015). Superabsorbent hydrogels based on polysaccharides for application in agriculture as soil conditioner and nutrient carrier: a review. *European Polymer Journal* 72: 365–385.

9 McClements, D.J. and Decker, E.A. (2017). Lipids. In: *Fennema's Food Chemistry*, 5e (eds. S. Damodaran and K.L. Parkin), 171–233. Boca Raton: CRC Press, Taylor & Francis Group.

10 Cropper, S.L., Kocaoglu-Vurma, N.A., Tharp, B.W., and Harper, W.J. (2013). Effects of locust bean gum and mono- and diglyceride concentrations on particle size and melting rates of ice cream. *Journal of Food Science* 78 (6): C811–C816.

11 Rubel, I.A., Iraporda, C., Gallo, A. et al. (2019). Spreadable ricotta cheese with hydrocolloids: effect on physicochemical and rheological properties. *International Dairy Journal* 94: 7–15.

5.8 Suggested Reading

Belitz, H.-D., Grosch, W., and Schieberle, P. (2009). *Food Chemistry*, 4e. Berlin: Springer. 1070 p.

Dziezak, J.D. (1991). A focus on gums. *Food Technology* 45 (3): 116–132.

Phillips, G.O. and Williams, P.A. (eds.) (2009). *Handbook of Hydrocolloids*, 2e. Cambridge: Woodhead Publishing. 924 p.

6

Functional Properties of Proteins

6.1 Learning Outcomes

After completing this exercise, students will be able to:

1) Assess the effects of pH on protein solubility.
2) Describe the role of divalent calcium ions in the gelation of soy proteins in solution.
3) Prepare soy protein extracts from basic materials such as defatted soy flour.
4) Produce soy-based products (i.e. tofu and soy protein isolate) in laboratory settings.
5) Use a dye-binding assay to measure protein concentration in solution.

6.2 Introduction

Proteins are an intrinsic component of most foods. They are also added in purified form during processing to perform certain desirable functions. From the perspective of a food scientist, the functional properties of proteins are those that influence food quality and appeal. These properties are determined by the primary, secondary, tertiary, and quaternary structures of proteins and vary widely among proteins. Heating, changes in pH, whipping, drying, and other treatments may alter the functional properties of proteins.

Some functions of proteins in foods include gelation, thickening, foaming, water holding, emulsification, coloring (e.g. myoglobin, Maillard browning products), cohesion, dough formation, and texturization. Thus, proteins are highly versatile food ingredients and are used in a wide range of products.

Protein functionality is difficult to study in complex food systems. Therefore, much of the research on protein functionality has been done in model systems where purified proteins are studied under carefully defined conditions.

Biochemists have invested years of research into developing techniques for purifying, identifying, quantifying, and characterizing proteins. Food scientists have applied many of these techniques to better understand the functions of proteins in foods and to develop processes for concentrating and isolating proteins for food use. For example, processes for producing high-quality food-grade soy protein concentrates and isolates from soybeans are widely used in the food industry. Moreover, results of protein research have helped us to better understand some traditional food processing operations that depend on the functional properties of proteins. Cheese making and tofu manufacturing are two prominent examples.

Food Chemistry: A Laboratory Manual, Second Edition. Dennis D. Miller and C. K. Yeung.
© 2022 John Wiley & Sons, Inc. Published 2022 by John Wiley & Sons, Inc.
Companion website: www.wiley.com/go/Miller/foodchemistry2

A key property of proteins is solubility. Many of the functions of proteins depend on solubility. For example, proteins must be in solution in order to form gels. Protein solubility is affected by pH, ionic strength, divalent cations, temperature, and the amino acid composition and sequence of the protein.

Proteins are large molecules that have strong tendencies to interact with other protein molecules, metal ions, water molecules, lipids, and carbohydrates. Thus, proteins can have marked and sometimes unpredictable effects on foods.

In this exercise, we will study the effects of pH and calcium ions on protein solubility. At low pH, soy protein has a net positive charge due to protonation of the carboxylate and amino groups. At high pH, soy protein has a net negative charge because of deprotonation of these same groups. When the proteins have a net positive or net negative charge, the individual molecules have a tendency to repel each other. This repulsion increases their solubility because they are more likely to interact with water than with other protein molecules. At the isoelectric point (pI), defined as the pH at which the net charge on a protein is zero, the number of positively charged side chains (e.g. protonated lysine residues) is equal to the number of negatively charged side chains (e.g. aspartate residues). Without a net positive or negative charge, protein molecules may associate, forming gels or precipitates. Thus, it is often possible to either solubilize protein powders or precipitate or gel dissolved protein by simply manipulating the pH. Cheese making relies, in part, on pH manipulation to form a protein gel.

Protein–protein associations may also be enhanced by adding multivalent ions. This approach is used in the production of tofu. This age-old product is made by solubilizing soy protein in water, heating to denature the protein to expose charged amino acid residues, and adding a calcium salt to cross-link the protein molecules.

6.2.1 Soybean Processing: Soy Milk, Tofu, and Soybean Protein Isolate

Soybeans have been consumed by humans for centuries. Raw mature soybean seeds are rich sources of protein (36 g/100 g), fat (20 g/100 g mostly polyunsaturated), fiber (9 g/100 g), calcium (277 mg/100 g), and iron (15 mg/100 g) among other nutrients [1]. The United States is the first or second leading producer of soybeans globally, but most of the soybeans grown in the United States are defatted and fed to animals. Soybean oil, produced by defatting whole soybeans, is the most widely consumed vegetable oil in the United States and is second only to palm oil globally [2].

Soy milk and tofu are two widely consumed manufactured foods made from soybeans. Soy milk is becoming increasingly popular in the United States as consumers are looking for alternatives to dairy milk. It is made by soaking whole mature soybeans in water, grinding the soaked beans into a slurry, boiling the slurry to denature proteins and deactivate trypsin inhibitors, and filtering or centrifuging to remove fibrous material called okara [3]. Tofu is made by coagulating the protein in soy milk. The protein can be coagulated by adding salts of divalent cations. Calcium sulfate (gypsum) or a mixture of calcium chloride and magnesium chloride (nigari) are commonly used. The protein can also be coagulated by acidifying the soy milk. Glucono-delta-lactone is a commonly used acidulant in the manufacture of tofu. After coagulation, the tofu is pressed to remove water, and the tofu is cut into pieces for packaging.

Another important soybean product is soy protein isolate. It is used as an ingredient in many manufactured foods. Soy protein isolate contains over 90% protein. Soy protein isolate is manufactured from dehulled, defatted, edible-grade soybean flakes (or flour). Protein is extracted from the flakes with an aqueous alkaline solution with a pH slightly higher than 7. Most proteins in soybeans are soluble at alkaline pHs, so the extract contains protein, soluble sugars and oligosac-charides, and other substances, leaving behind most polysaccharides. Once the proteins have been solubilized, the extract is clarified by centrifugation. The supernatant is then acidified to pH 4–5, which is within the isoelectric range for most of the proteins. This causes the proteins to precipitate into a curd leaving the sugars and oligosaccharides in solution. The supernatant is removed and the curd is spray dried to yield a soy protein isolate powder. The curd may be neutralized to pH 6.5–7.0 prior to drying [4].

6.2.2 Assaying Protein Concentration

Several assays are available for determining the concentration of protein in solution. One of the simplest of these assays is the so-called dye-binding assay developed by Bradford [5, 6]. As the name implies, a dye added to a protein solution binds to the available protein producing a dye–protein complex. The acidic dye Coomassie® Brilliant Blue G-250 is used. It has an absorbance maximum of 465 nm when free in solution. When it binds to protein, however, the absorbance maximum shifts to 595 nm. Thus, the absorbance at 595 nm is proportional to the concentration of the protein.

Protein concentration in samples is determined by comparing the absorbance of samples to a standard curve. Unfortunately, the extent of dye binding varies from protein to protein since the dye binds primarily to basic and aromatic amino acids and the proportion of these amino acids varies from protein to protein. Thus, unless the standard is made from the protein being measured, the assay provides a relative rather than an absolute measure of the protein present in the sample. Bovine serum albumin (BSA) is routinely used for the standard because it readily dissolves in water and it is available in purified form.

Several laboratory supply companies have developed kits for this assay, making the assay even simpler. A kit supplied by Bio-Rad Laboratories called the Bio-Rad Protein Assay is widely used. For details on the principles of dye-binding protein assays and a protocol for measuring protein concentrations using the assay, see the booklet titled Bio-Rad Protein Assay [7].

6.3 Apparatus and Instruments

1) Heat/stir plate and magnetic stir bars
2) Spectrophotometer and cuvettes
3) Centrifuge
4) pH meter and pH standards
5) Electronic top-loading balance
6) Centrifuge tubes with caps (10 per group)
7) Plastic centrifuge tubes with caps (50 ml capacity)
8) Beakers, 25 and 50 ml
9) Graduated cylinders
10) Test tubes (20 per group)
11) Pipettes
12) Pasteur pipettes & bulbs
13) Funnel
14) Filter paper
15) Parafilm®

6.4 Reagents and Materials

1) Defatted soy flour (10 g per group)
2) NaOH, 0.1 N
3) HCl, 1.0 N
4) $CaCl_2$, 5 M
5) Bradford reagent (0.004% Brilliant Blue G in 10% phosphoric acid/4% methanol. Handle with gloves and eye protection. Dispose in waste bottle).
6) Bovine serum albumin (BSA) – 1.0 mg ml^{-1} in H_2O

6.5 Procedures

Note: Complete instructions and background information for the Bio-Rad Bradford Protein assay are available online at: https://www.bio-rad.com/webroot/web/pdf/lsr/literature/LIT33.pdf

6.5.1 Standard Curve for the Bradford Protein Assay

1) Prepare a series of standard solutions of BSA in water consisting of the following concentrations: 0.25, 0.5, 0.75, and 1.0 mg ml^{-1}. Prepare 1 ml of each solution by making appropriate dilutions of the 1.0 mg ml^{-1} BSA solution.
2) Transfer, in duplicate, 50 µl of each standard to clean, dry test tubes.
3) Add 2.5 ml of the Bio-Rad dye reagent to each tube.
4) Cover tubes with Parafilm® and mix by inverting several times.
5) Incubate at room temperature for at least five minutes but no longer than one hour.
6) Read absorbance at 595 nm against the reagent blank, i.e. zero the spectrophotometer with the reagent blank. (The reagent blank should be identical to the standards except that it should contain no protein.)
7) Plot a standard curve (absorbance versus BSA concentration).

6.5.2 Effect of pH on Protein Solubility

6.5.2.1 Preparation of Protein Extracts
Note: Proteins will be extracted from soy flour dispersed in water and adjusted to a range of pH values (2.5, 3.0, 3.5, 4.0, 4.5, 5.0, 5.5, 6.0, 6.5, 7.0, 7.5, 8.0, and 8.5). Each group will extract at one (or two) pH.

1) Prepare a 1 : 15 soy flour/water mixture by adding 10 g of soy flour to 150 ml distilled H_2O and stirring on a magnetic stir plate to form a uniform suspension.
2) Transfer a 15 ml aliquot of the suspension to a 50 ml beaker (remove the aliquot with the stirrer running to prevent insolubles from settling).
3) Adjust to your assigned pH with 1.0 N HCl or 0.1 N NaOH (you will be assigned a pH at the beginning of class). Add distilled water to bring total volume to 20 ml (the graduations on the beaker are sufficient for this volume adjustment).
4) Stir on a stir plate for 20 minutes at room temperature.
5) Transfer 10 ml to a screw-capped centrifuge tube and centrifuge at full speed for 10 minutes in a bench-top centrifuge. Be sure you balance your tubes.
6) Decant supernatant into a clean tube.

6.5.2.2   Measurement of Protein Concentration in the Extracts

1) Make a 1 to 50 dilution of the supernatant from the protein extraction above (0.1 ml supernatant + 4.9 ml dH$_2$O) for the samples with pH values of below 4.0 and above 5.0. For pH values of 4.0, 4.5, and 5.0, make a 1 to 10 dilution.
2) Determine the protein concentration in the diluted supernatant using the Bradford protein assay and the standard curve you constructed in Section 6.5.1. (Read against a reagent blank as you did for the standard curve). Follow the same protocol you used in preparing the standard curve (50 µl sample, 2.5 ml Bio-Rad dye reagent). **Note**: Your absorbance reading should be greater than 0.1. If it is less than 0.1, your sample is too dilute and you should take another aliquot of the supernatant, dilute it less, and re-run the assay.
3) Calculate the amount (in g) of soluble protein in the soy flour extract. Be sure to account for the dilutions you made.
4) Calculate the solubility of the protein in the soy flour (express as a percentage of the protein in the flour). Assume the soy flour is 50% protein. Remember that a 15 ml aliquot of the original mixture contains 1 g soy flour.
5) Using your data and data from the rest of the class, construct a plot of protein solubility versus pH.

6.5.3   Preparation of Soy Protein Isolate and Tofu

6.5.3.1   Extraction

1) Transfer 100 ml of the 1 : 15 soy flour/water mixture from Step 1 in Section 6.5.2.1 to a beaker (be sure to mix well before transferring). Stir on a magnetic stirrer.
2) Adjust to pH 8 with 0.1 N NaOH.
3) Stir at this pH for 20 minutes at room temperature.
4) Transfer the suspension to two 50-ml centrifuge tubes. Centrifuge at approximately 10,000 × g for 20 minutes. Pool the supernatants into one beaker.
5) Determine the protein concentration in the supernatant.
6) Divide the pooled supernatants into two 25 ml aliquots in 50 ml beakers.

6.5.3.2   Soy Protein Isolation

1) Measure the pH of one of the 25 ml aliquots of the above extract (it should be near pH 8).
2) Add one drop of 1.0 N HCl *without stirring* and record what happens.
3) Begin stirring on a stir plate and adjust to pH 4.5 with 1.0 N HCl. Record your observations.
4) Filter the mixture from Step 3 and collect the filtrate. The residue on the filter paper is the soy protein isolate.
5) Determine the protein concentration of the filtrate with the dye-binding assay (remember to dilute before measuring the protein concentration).
6) Calculate protein yield in the isolate (the percentage of the protein in the flour that was recovered in the isolate). Note that you are measuring the amount of protein that was not recovered, so you will determine recovered protein by difference.

6.5.3.3   Production of Tofu

1) Stir the other 25 ml aliquot from Section 6.5.3.1 and heat to boiling. (Remove from the heat as soon as boiling begins.)
2) Add a drop of 5 M CaCl$_2$ to the extract while it is still hot and record what happens.

3) Add a few more drops and stir with a glass stirring rod.
4) Pick out a clump of the "tofu" and feel the texture. (Compare the texture with the pH 4.5 isolate.)

6.6 Problem Set

1 A food chemistry student completed the work outlined above and obtained the following data:
Standard curve: Absorbances at 595 nm for the BSA standards: 0.20, 0.43, 0.65, and 0.90 for 0.25, 0.5, 0.75, and 1.0 mg BSA ml^{-1}, respectively.
Supernatant absorbance: The absorbance in the dye-binding assay for the diluted supernatant was 0.35.
   a) Construct a standard curve (absorbance versus protein concentration in the standards).
   b) Calculate the concentration of protein in the diluted supernatant and the supernatant prior to dilution.
   c) Calculate the amount of soluble protein (in g) in the extract.
   d) Calculate the solubility of the protein (express as a percentage of the protein in the soy flour). Assume the soy flour was 50% protein (w/w).

6.7 Study Questions

1 Draw the structure of the following peptide: ala-asp-gly-lys-ser-glu-val-arg-his-gly. What would the net charge on this peptide be at pH 2.0? At pH 8.0? (See [8] or a biochemistry textbook for amino acid structures and pK_a values).

2 Explain, showing partial structures, how calcium ions cause soy protein to gel.

3 Why do proteins form precipitates under some conditions and gels under others?

6.8 References

1 USDA (2020). FoodData central [Internet]. https://fdc.nal.usda.gov/ (accessed 11 February 2020).
2 Statista (2020). Global vegetable oil consumption, 2018/19 [Internet]. https://www.statista.com/statistics/263937/vegetable-oils-global-consumption/ (accessed 11 February 2020).
3 McHugh, T. (2016). How tofu is processed. *Food Technology* 70 (2): 72–74.
4 Lusas, E.W. and Riaz, M.N. (1995). Soy protein products: processing and use. *The Journal of Nutrition* 125 (suppl_3): 573S–580S.
5 Bradford, M.M. (1976). A rapid and sensitive method for the quantitation of microgram quantities of protein utilizing the principle of protein-dye binding. *Analytical Biochemistry* 72 (1–2): 248–254.
6 Kruger, N.J. (1994). The Bradford method for protein quantification. In: *Basic protein and peptide protocols* (ed. J.M. Walker), 9–16. Totowa, N.J: Humana Press. (Methods in molecular biology).
7 Bio-Rad (2020). Bio-Rad protein assay | Life Science Research | Bio-Rad [Internet]. https://www.bio-rad.com/en-us/product/bio-rad-protein-assay?ID=d4d4169a-12e8-4819-8b3e-ccab019c6e13 (accessed 11 February 2020).
8 Damodaran, S. (2017). Amino acids, peptides, and proteins. In: *Fennema's Food Chemistry*, 5e (eds. S. Damodaran and K.L. Parkin), 235–356. Boca Raton: CRC Press, Taylor & Francis Group.

6.9 Suggested Reading

Fukushima, D. (1991). Recent progress of soybean protein foods: chemistry, technology, and nutrition. *Food Reviews International* 7 (3): 323–351.

Smith, A.K. and Circle, S.J. (1978). *Soybeans: Chemistry and Technology (Volume I - Proteins)*. Westport, CT: Avi Publishing Company,Inc.

Answers to Problem Set

1 a) Equation for standard curve: $y = 0.928x - 0.035$; $R^2 = 0.9993$
   b) $0.415\,mg\,ml^{-1}$ (diluted); $27.67\,mg\,ml^{-1}$ (undiluted)
   c) $4.15\,g$
   d) 83%

7

Lactose

7.1 Learning Outcomes

After completing this exercise, students will be able to:

1) Explain what lactose intolerance is and why it affects some people and not others.
2) Design an experiment to test the hypothesis that live cultures in yogurt aid in the digestion of lactose in the small intestine.
3) Measure the lactose content of fluid milk and yogurt samples.
4) Measure the lactase activity in regular and Greek yogurts under simulated intestinal conditions.

7.2 Introduction

Lactose (milk sugar) is a disaccharide composed of galactose and glucose linked by a β-(1,4)-galactosidic bond. As shown in Figure 7.1, glucose provides the alcohol group and galactose provides the glycosyl group to form the disaccharide. According to convention, the glucose is called an aglycon when it is part of lactose.

Lactose cannot be absorbed intact from the gastrointestinal tract. It must first be hydrolyzed to glucose and galactose, a reaction that is very slow unless catalyzed by lactase, a β-galactosidase (Figure 7.2). Most mammals, with the exception of humans whose ancestry includes northern Europeans and some populations from Africa and India, lose the ability to digest lactose after they reach weaning age. It has been estimated that the prevalence of lactase deficiency in non-white adults is as high as 50–75%. When unabsorbed lactose passes into the colon, it may draw water into the lumen of the colon causing diarrhea or be fermented by the colonic microflora producing gas, which causes bloating and flatulence. People who exhibit these symptoms following consumption of dairy products are commonly diagnosed as lactose intolerant. People who are lactose intolerant may avoid dairy products since their consumption may cause unpleasant gastrointestinal symptoms.

The principal dietary source of lactose in the United States is bovine milk, which contains slightly less than 5% (w/w) of the sugar. Other dairy products contain varying amounts of lactose, depending on processing, and storage conditions.

Fermentation has been used for centuries as a method for preserving milk. Cheeses and other fermented dairy products are well known and widely consumed. Bacterial cultures used for these

Food Chemistry: A Laboratory Manual, Second Edition. Dennis D. Miller and C. K. Yeung.
© 2022 John Wiley & Sons, Inc. Published 2022 by John Wiley & Sons, Inc.
Companion website: www.wiley.com/go/Miller/foodchemistry2

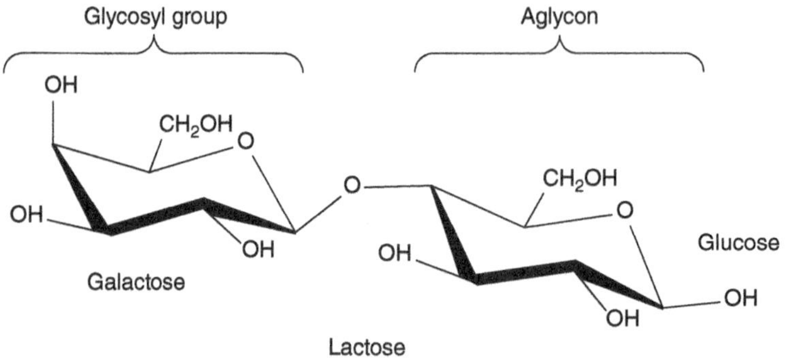

Figure 7.1 Lactose (conformational representation) is a disaccharide composed of galactose and glucose linked by a β-(1,4)-galactosidic bond.

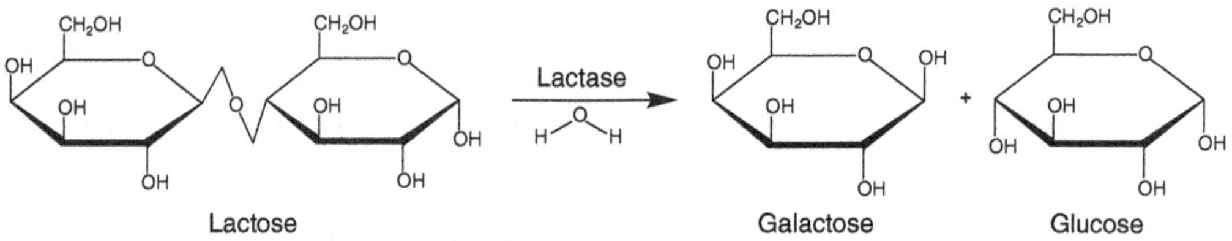

Figure 7.2 Hydrolysis of lactose to D-galactose and D-glucose (Haworth representation). This reaction is catalyzed by lactase (β-galactosidase).

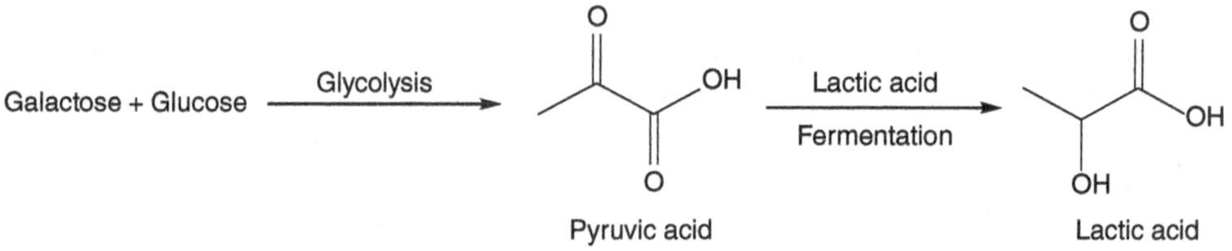

Figure 7.3 Bacterial cultures in fermented dairy products convert galactose and glucose into pyruvic acid via the glycolytic pathway. Pyruvic acid undergoes lactic acid fermentation to produce lactic acid, which lowers the pH of the product.

fermentations produce β-galactosidases, which can significantly reduce the lactose content of the product through hydrolysis and subsequent lactic acid fermentation (Figure 7.3).

Hard cheeses generally contain very little lactose and are well tolerated by lactose-intolerant individuals. Yogurts, on the other hand, may contain concentrations of lactose similar to or only slightly lower than that in milk since most yogurt mixes are formulated to contain nonfat dry milk powder in addition to fluid milk. Nonfat milk powder contains approximately 50% lactose (w/w). Moreover, lactase activity is pH sensitive and decreases as the pH falls during fermentation. This further explains why lactose concentrations in yogurt are relatively high. Even so, many lactose-intolerant individuals claim they can consume yogurt without the symptoms they experience when drinking milk. This has led to the hypothesis that the enzymes produced by the cultures used in yogurt manufacture (*Lactobacillus bulgaricus* and *Streptococcus thermophilus*) may still be active in the guts of yogurt consumers and may aid in the digestion of lactose. Kolars et al. [1] have reported convincing evidence in support of this hypothesis.

A more recent approach to lactose reduction in dairy foods is the addition of lactase to the product by either the processer or the consumer. LACTAID® is a commercially available preparation of lactase that may be added to milk and other dairy products during processing or ingested by consumers as a dietary supplement. Manufacturers recommend that lactose intolerant individuals take a LACTAID® supplement (available as chewables or caplets) just before consuming a lactose-containing product (https://www.lactaid.com/). The fact that consuming LACTAID® prior to eating a dairy product prevents symptoms is a strong evidence that ingested lactase survives the acid stomach and is active in the small intestine.

7.2.1 Lactose Assay

Concentrations of many compounds in solution may be determined by measuring the absorbance of the solution with a spectrophotometer. However, this works only if the compound in question has a unique absorption maximum, i.e. it must absorb at a wavelength at which no other compound in the solution absorbs. Lactose does not have a distinct absorption maximum and therefore cannot be measured directly by spectrophotometry. Fortunately, lactose and many other compounds can be measured indirectly by taking advantage of coupled reactions with other compounds that do have distinct absorption maxima. A coupled reaction with NAD^+ or $NADP^+$ is commonly used in biochemical assays. Reactions of NAD^+ or $NADP^+$ with various substrates are enzyme catalyzed and thus allow for highly specific assays if a purified form of a suitable enzyme is available. NAD^+ and $NADH^+$ are oxidizing agents and react stoichiometrically with reduced substrates (such as reducing sugars) to produce an oxidized substrate (e.g. galactonic acid from the oxidation of galactose) and NADH or NADPH. NAD^+ (or $NADP^+$) and NADH (or NADPH) have distinctly different absorption spectra (NADH absorbs at 340 nm, whereas NAD^+ does not) (Figure 7.4). Therefore, the conversion of NAD^+ to NADH in a solution can be monitored by measuring the absorbance at 340 nm.

Commercial kits are available for measuring lactose in various dairy products. Megazyme (Bray, Ireland) markets one such kit. Detailed instructions for using the Megazyme kit to measure lactose in milk and yogurt are described in a booklet authored by Megazyme [3].

The first step in the enzymatic assay for lactose is to hydrolyze the lactose to form glucose plus galactose. This is accomplished by adding lactase to the sample containing the lactose (Figure 7.2).

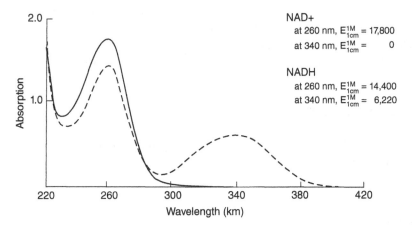

Figure 7.4 Absorption spectra of NAD^+ (solid line) and NADH (dashed line). *Source:* From [2].

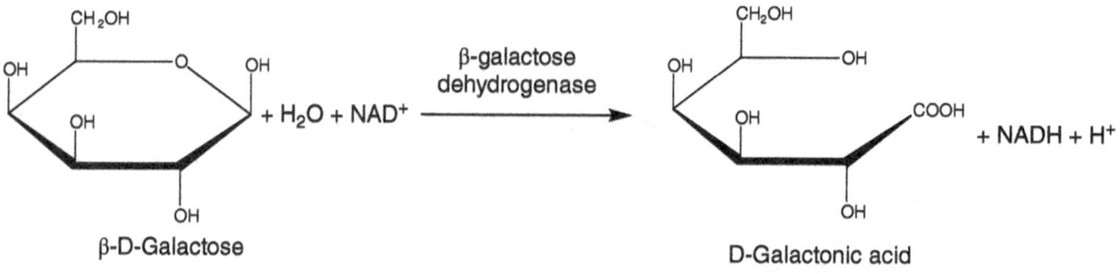

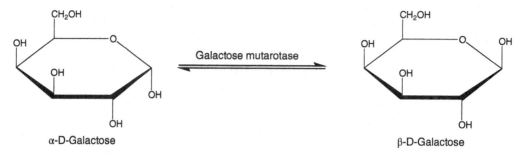

Figure 7.5 The oxidation of β-D-galactose by NAD$^+$, generating NADH. The reaction is catalyzed by β-galactose dehydrogenase.

CH$_2$OH

OH   O   OH

OH

     Galactose mutarotase

OH

OH

α-D-Galactose

CH$_2$OH

OH   O   OH

OH

OH

β-D-Galactose

Figure 7.6 The equilibrium of α- and β-D-galactose in solution. The attainment of the equilibrium is rather slow. In the presence of the enzyme galactose mutarotase; however, the conversion is quite rapid and for that reason, it is often used in lactose assays.

The next step is to oxidize the galactose to D-galactonic acid using NAD$^+$ as the oxidizing agent and β-galactose dehydrogenase as a catalyst as shown in Figure 7.5. Note that the number of moles of NADH produced is stoichiometric with the number of moles of galactose, which is stoichiometric with the number of moles of lactose in the sample.

Note that β-galactose dehydrogenase is specific for β-D-galactose. It does not catalyze the oxidation of α-D-galactose. Galactose, like lactose and glucose, is a reducing sugar. Therefore, it exists in solution in two forms in equilibrium: α-D-galactose and β-D-galactose (Figure 7.6). As the β-galactose is oxidized, the equilibrium shifts, generating more β-D-galactose, but this conversion is rather slow and results in a "creeping" reaction, i.e. the absorbance at 340 nm increases gradually over time. Thus, it takes a long time (several hours) for the complete oxidation of all of the galactose in the solution. This makes for a less than ideal assay. Fortunately, there is an enzyme that dramatically speeds up this conversion. It is called galactose mutarotase (Figure 7.6).

7.3 Apparatus and Instruments

The equipment you will need for this exercise will vary somewhat depending on the assay kits you choose. Most will require the following:

1) UV-spectrophotometer or micro-plate reader
2) Water bath
3) Cuvettes (or micro-plates)
4) Test tubes

5) Vortex mixer
6) Timer

7.4 Reagents and Materials

1) Pasteurized skim milk
2) Pasteurized LACTAID® skim milk
3) Plain low-fat (or nonfat) yogurt
4) Plain low-fat yogurt Greek style
5) Commercial assay kit for measuring concentrations of lactose and galactose in dairy products. Suitable kit:
   - Megazyme: Lactose Assay Kit – Sequential/High Sensitivity [3]
6) Commercial assay kit for measuring beta-galactosidase activity in yogurts. Suitable kit:
   - ThermoFisher Scientific: Yeast beta-Galactosidase Assay Kit [4]

7.5 Procedures

7.5.1 Lactose and D-galactose Assay Protocol

The protocol for this assay will vary somewhat depending on the supplier of the kit. Follow the directions that come with the kit you will use for this assay.

7.5.2 Lactase Assay

A common method for measuring lactase activity is to incubate a lactase-containing food or tissue with a compound that acts as a substrate for the enzyme. One such compound is o-nitrophenyl-β-D-galactopyranoside (ONPG). Lactase catalyzes the hydrolysis of ONPG to o-nitrophenol and galactose. ONPG is colorless, but o-nitrophenol is yellow and can be quantified in solution by measuring the absorbance at 420 nm. ThermoFisher Scientific (Waltham, MA) offers a kit using ONPG for measuring β-galactosidase in yeast. It will also work for measuring lactase activity in yogurt.

7.6 Experimental Design

For this experiment, each student or lab group should develop their own experimental design for determining the concentration of lactose in various dairy products and for testing the hypothesis that live cultures in yogurt aid in the digestion of lactose in the small intestine. Recall that the pH of the small intestine is about 7, while the pH of many yogurts is below 5. To test this hypothesis, you will need to adjust the pH of the yogurt to around 7 and incubate at 37 °C for 30 minutes to one hour. Acids in digesta that enter our small intestines from the stomach are neutralized by pancreatic juice. The base in pancreatic juice is sodium bicarbonate, so we recommend using a solution of sodium bicarbonate ($0.5 \, \text{mol} \, \text{l}^{-1}$ should work) to raise the pH of the yogurt to 7. We suggest using a plain yogurt with live and active cultures and without added sugar. Other yogurts (e.g. Greek style or drinkable) may be used as well.

7.7   Study Questions

1   Compare the lactose content of all of your samples. How do they compare with literature values for similar samples?

2   Did adjusting the pH of your yogurt samples to 7 and incubating affect the lactose content of the yogurt?

3   In addition to the live bacteria present in yogurt, what other attributes of yogurt might help the digestion of lactose in the gastrointestinal environment?

4   Would you recommend to a friend who is lactose intolerant and who avoids all dairy products that she should try eating yogurt?

7.8   References

1 Kolars, J.C., Levitt, M.D., Aouji, M., and Savaiano, D.A. (1984). Yogurt—an autodigesting source of lactose. *New England Journal of Medicine* 310 (1): 1–3.
2 Switzer, R.L. and Garrity, L.F. (1999). *Experimental Biochemistry*, 3e. New York: W. H. Freeman and Co. 451 p.
3 Megazyme (2020). Lactose assay kit [Internet]. https://www.megazyme.com/lactose-assay-kit (accessed 11 February 2020).
4 ThermoFisher (2020). Yeast beta-galactosidase assay kit [Internet]. http://www.thermofisher.com/order/catalog/product/75768 (accessed 11 February 2020).

7.9   Suggested Reading

Chandan, R.C. (2017). An overview on yogurt production and composition. In: *Yogurt in Health and Disease Prevention* (ed. N.P. Shah), 31–47. London: Academic Press, an imprint of Elsevier.
Lu, Y., Ishikawa, H., Kwon, Y. et al. (2018). Real-time monitoring of chemical changes in three kinds of fermented milk products during fermentation using quantitative difference nuclear magnetic resonance spectroscopy. *Journal of Agricultural and Food Chemistry* 66 (6): 1479–1487.
Lynch, J.M., Barbano, D.M., and Fleming, J.R. (2007). Determination of the lactose content of fluid milk by spectrophotometric enzymatic analysis using weight additions and path length adjustment: collaborative study. *Journal of AOAC International* 90 (1): 196–216.
Pochart, P., Dewit, O., Desjeux, J.F., and Bourlioux, P. (1989). Viable starter culture, beta-galactosidase activity, and lactose in duodenum after yogurt ingestion in lactase-deficient humans. *The American Journal of Clinical Nutrition* 49 (5): 828–831.
Savaiano, D.A. (2014). Lactose digestion from yogurt: mechanism and relevance. *The American Journal of Clinical Nutrition* 99 (5): 1251S–1255S.
Zimmerman, T., Ibrahim, M., Gyawali, R., and Ibrahim, S.A. (2019). Linking biochemistry concepts to food safety using yogurt as a model. *Journal of Food Science Education* 18 (1): 4–10.

8

Enzymatic Browning: Kinetics of Polyphenoloxidase

8.1 Learning Outcomes

After completing this exercise, students will be able to:

1) Prepare a crude enzyme extract from a natural source.
2) Perform an enzyme assay.
3) Determine K_M and V_{max} for an enzyme using data obtained in an enzyme assay.

8.2 Introduction

The browning that develops when cut or bruised surfaces of fruits, vegetables, and shellfish are exposed to air is called enzymatic browning because the initial reactions involved are enzyme catalyzed [1]. The enzyme responsible for the initiation of this browning reaction has several common names including phenolase, phenoloxidase, tyrosinase, polyphenoloxidase (PPO), and catecholase [2]. These oxidases are present in both plant and animal tissues. In animals, the enzyme is usually called tyrosinase (because tyrosine is one of its substrates). An important function of tyrosinase is to catalyze the formation of brown melanin pigments, which impart color to skin, hair, and eyes. In plants, the enzyme is more commonly called PPO, suggesting that its primary substrates are polyphenolic compounds. The function of the enzyme in plants is unknown, but it is responsible for significant color changes (both beneficial and detrimental) in many foods. In intact plant tissue, PPO and its phenolic substrates are separated by cell structures and browning does not occur. Cutting, bruising, or otherwise damaging the integrity of plant tissues often allows the enzyme and its substrate to come into contact. A generalized reaction sequence for enzymatic browning is shown in Figure 8.1.

Common substrates for PPO in plant tissues include the amino acid tyrosine and polyphenolic compound such as catechin, caffeic acid, and chlorogenic acid (Figure 8.2). Tyrosine, being a monophenol, is first hydroxylated to 3,4-dihydroxyphenylalanine (dopa) and then is oxidized to a quinone (Figure 8.3).

8.2.1 Enzyme Kinetics

Please refer to your food chemistry or biochemistry textbook for a detailed discussion of enzyme kinetics.

Food Chemistry: A Laboratory Manual, Second Edition. Dennis D. Miller and C. K. Yeung.
© 2022 John Wiley & Sons, Inc. Published 2022 by John Wiley & Sons, Inc.
Companion website: www.wiley.com/go/Miller/foodchemistry2

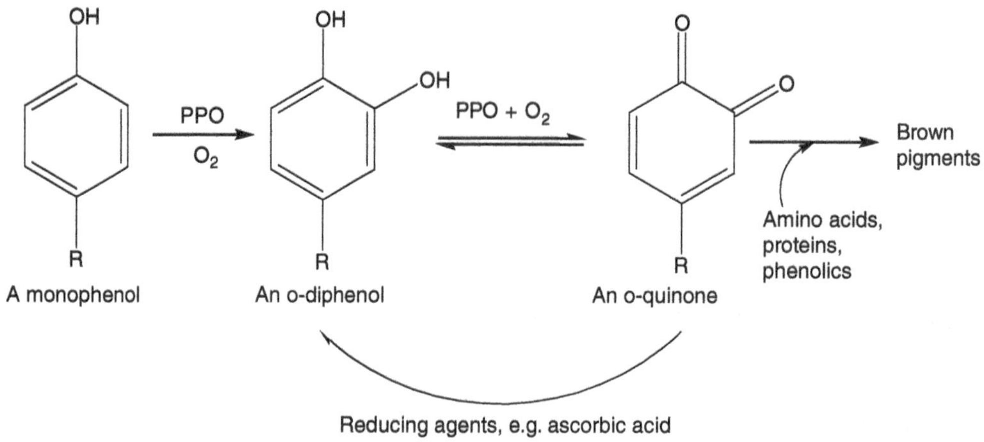

Figure 8.1 The action of polyphenoloxidase (PPO) on phenolic compounds. *Source:* Adapted from [3].

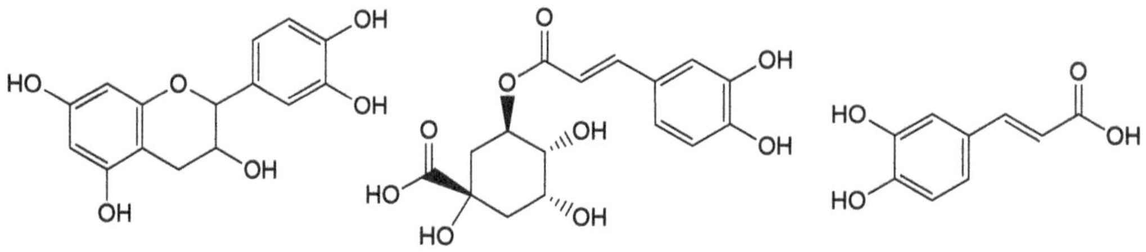

Catechin   Chlorogenic acid   Caffeic acid

Figure 8.2 Examples of PPO substrates found in foods. Note that they are polyphenols.

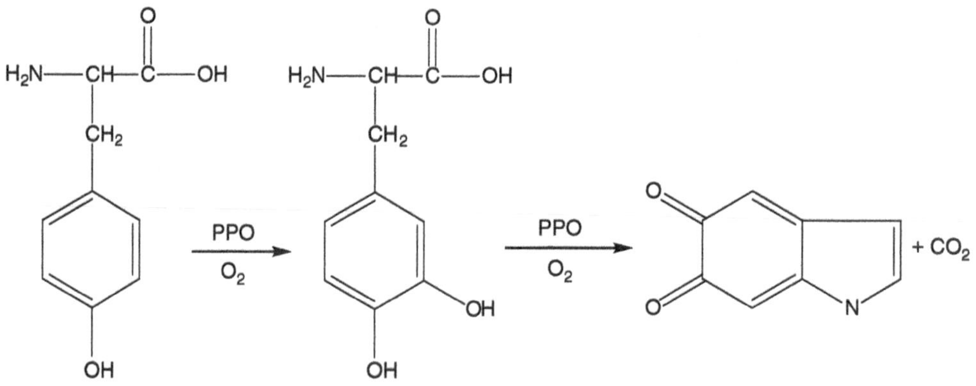

Tyrosine   Dopa (3,4-dihydroxyphenylalanine)   Indol-5,6-quinone

Figure 8.3 The action of PPO on tyrosine to produce indol-5,6-quinone [4].

The rates of reactions catalyzed by specific enzymes are called enzyme activities. Enzyme activities may be determined by measuring the rate of disappearance of substrates or the rate of appearance of products. PPO, therefore, could conceivably be assayed by measuring oxygen uptake, disappearance of phenolic compounds, brown color formation, or the formation of an intermediate in the reaction such as a specific quinone. Indole-5,6-quinone, a product of the PPO-catalyzed oxidation of 3,4-dihydroxyphenylalanine, is a convenient product to follow since it absorbs light at a wavelength where other substances present do not interfere.

In many situations, the actual concentration of an enzyme in a tissue extract is unknown. Therefore, enzyme activity rather than concentration or weight is used to quantify the amount present. Activity is usually expressed as the rate or velocity of the reaction catalyzed by the enzyme. Rate and velocity are really the same thing but are expressed in different units (see below). Enzyme activities are determined with enzyme assays where the enzyme and its substrate(s) are mixed together and the progress of the reaction is monitored over time. When the substrate is present in excess and temperature and pH are properly controlled, velocity is proportional to enzyme concentration, i.e. the higher the enzyme concentration, the faster the reaction.

In this experiment, rate is defined as the change in absorbance of the assay mixture per unit time, i.e. it is equal to the slope of the linear portion of a plot of absorbance versus time (Figure 8.4).

$$\text{Rate} = \frac{\Delta A}{\Delta t}$$

It is important to measure the initial rate in an enzyme assay because the rate will slow as the reaction proceeds. This is because reaction rate is dependent on substrate concentration and substrate concentration decreases with time. Also, in some reactions, product inhibition will slow the reaction.

Velocity is defined at the change in concentration of a reactant or product with time. In the case where dopa is oxidized in a reaction catalyzed by PPO, absorbance (A) is directly proportional to the concentration of indole-5,6-quinone (IQ), a product of the reaction. Thus, we can calculate velocity from the rate if we know the relationship between absorbance and concentration. Absorbance is related to concentration by the Beer-Lampert Law:

$$A = \varepsilon bc$$

where

ε = absorbancy coefficient (5.0121 mmol^{-1} cm^{-1} for IQ)
b = length of light path (1.0 cm)
c = concentration (mmol l^{-1})

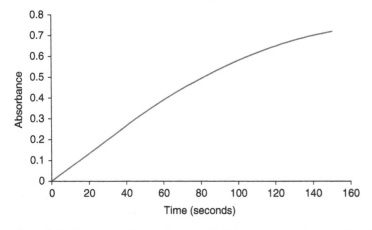

Figure 8.4 The change in absorbance with time in an enzyme assay. The assay mixture contains the enzyme and its substrate. Absorbance is a measure of product concentration, which increases with time. The initial rate of the reaction is equal to the slope of the linear portion of the line.

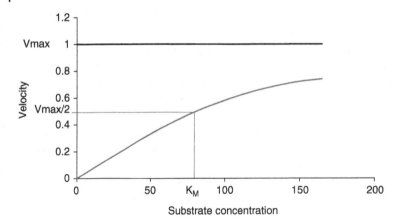

Figure 8.5 A plot of velocity versus substrate concentration for an enzyme that obeys Michaelis–Menten kinetics.

We have defined velocity as:

$$\text{Velocity}\left(V\right) = \frac{\Delta c}{\Delta t}\left(\text{mmol}\right)\left(l^{-1}\right)\left(\text{min}^{-1}\right)$$

Substituting with the formula for the Beer-Lampert law, we have:

$$V = \frac{\Delta A}{\varepsilon b \Delta t} = \frac{\text{rate}}{\varepsilon b} = \frac{\text{rate}}{5.012}\left(\text{mmol}\right)\left(l^{-1}\right)\left(\text{min}^{-1}\right)$$

The velocities of many enzyme catalyzed reactions exhibit a hyperbolic relationship to substrate concentration as shown in Figure 8.5. Enzymes that exhibit this relationship are said to obey Michaelis–Menten kinetics.

Notice that at low substrate concentrations, velocity is nearly linearly related to substrate concentration, while at high substrate concentrations velocity is almost unaffected by changes in substrate concentration. Michaelis and Menten proposed a model to explain this hyperbolic relationship. They suggested that an enzyme (E) forms a complex (ES) with a substrate (S) and that the substrate in the complex can either dissociate unchanged from the enzyme or it can be converted to a product (P) and then dissociate. Their model can be summarized as follows:

$$E + S \underset{k_2}{\overset{k_1}{\rightleftharpoons}} ES \xrightarrow{k_3} E + P$$

Velocity of the reaction shown in the above model is equal to a rate constant times the concentration of the enzyme–substrate complex:

$$V = k_3\left[ES\right]$$

Clearly, when the substrate concentration is high enough to saturate all active sites on the enzyme, further increases in substrate concentration will not affect velocity since at high substrate concentrations, the concentration of ES is limited by enzyme concentration, not substrate concentration.

V_{max} is the maximal rate attained when all the active sites of an enzyme are saturated with substrate. It reveals the turnover number of the enzyme, i.e. the number of substrate molecules converted to product per unit time.

Michaelis and Menten defined another constant, which they called K_M:

$$K_M = \frac{k_2 + k_3}{k_1}$$

K_M equals the concentration of substrate at which the reaction rate is half its maximal value. It is a measure of the affinity of the enzyme for its substrate. A large K_M indicates weak binding, and a small K_M strong binding.

K_M and V_{max} can be determined from reaction velocities measured at different substrate concentrations providing the enzyme follows Michaelis–Menten kinetics. The Michaelis–Menten equation describes the kinetics of enzymes, which fit the Michaelis–Menten model:

$$V = V_{max} \frac{[S]}{[S] + K_M}$$

The reciprocal of the Michaelis–Menten equation describes a straight line:

$$\frac{1}{V} = \frac{1}{V_{max}} + \frac{K_M}{V_{max}} \cdot \frac{1}{[S]}$$

Thus, we would expect a plot of 1/V versus 1/[S] to be a straight line. Figure 8.6 shows a plot of 1/V vs 1/[S]. It is commonly referred to as a Lineweaver-Burk plot. The y-intercept is $1/V_{max}$ and the slope is K_M/V_{max}. The x-intercept is $-1/K_M$.

8.2.2 PPO Assay

In this experiment, PPO will be extracted from potatoes and assayed using 3,4-dihydroxyphenylalanine (dopa) as the substrate. The progress of the reaction will be followed by monitoring the formation of indole-5,6-quinone, an early product of the oxidation. Indole-5,6-quinone absorbs strongly at 475 nm, thus absorbance at 475 nm will be used to quantitate the production of indole-5,6-quinone.

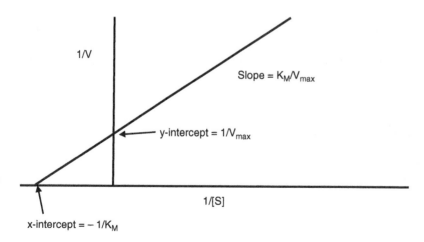

Figure 8.6 A Lineweaver-Burk plot for an enzyme that obeys Michaelis–Menten kinetics. V = velocity of the reaction; [S] = substrate concentration.

The assay will be set up to generate data that will allow calculation of V_{max} and K_M. In addition, we will study the effects of an inhibitor and of the enzyme concentration on the rate of the reaction.

8.2.3 Control of Enzymatic Browning

Enzymatic browning of foods is often, but not always, an adverse change because it reduces the acceptability of the food. Thus, a large body of literature on safe and effective methods for preventing enzymatic browning exists.

Three components must be present for enzymatic browning to occur: active PPO, oxygen, and a suitable substrate. Elimination of any of these will prevent the reaction from occurring. In addition, reducing agents capable of converting o-quinones back to phenolic compounds may effectively reduce browning.

Several different methods based on one or more of the above considerations have been used for the control of enzymatic browning in foods [3]. A brief summary of these methods follows:

1) Inactivation of PPO with heat. This approach is often used with vegetables that will ultimately be cooked before consumption. Heating to temperatures required for inactivation of PPO may not be suitable for fruits since it can impart undesirable cooked flavors and/or soft textures.
2) Chemical inhibition of PPO. Several strategies are used. Sulfites are extremely effective inhibitors of PPO, but the FDA restricts their use because they can cause life–threatening allergic reactions in some people. Acidulants such as citric acid inhibit the enzyme by lowering the pH below the optimum range. Chelating or sequestering agents such as EDTA and citric acid may inhibit the enzyme by binding copper, an essential cofactor.
3) Reducing agents. Agents that reduce o-quinones to phenolic compounds inhibit enzymatic browning. Ascorbic and erythorbic (D-isoascorbic) acids have been used to prevent browning of fresh-cut fruits for more than 50 years. Sulfites, in addition to directly inhibiting PPO, are also effective reducing agents but, again, FDA regulations limit their use in foods.
4) Exclusion of oxygen. Frozen peach slices are packed in evacuated, sealed containers to prevent exposure to oxygen. Sugar syrups are often used in fruit packs to exclude oxygen. Fruit pieces may be coated with an edible oxygen-impermeable edible film made from xanthan or other gums.
5) Proteolytic enzymes. Proteolytic enzymes that attack PPO and inactivate it have been proposed but are not yet widely used.
6) Treatment with honey. Honey apparently contains inhibitors of PPO, and research has shown it is effective on cut fruits. It is not widely used.

8.3 Apparatus and Instruments

1) Visible spectrophotometer
2) Pipettes, 2 and 5 ml
3) Pipettor, 250–1,000 μl
4) Cuvettes
5) Test tubes
6) Parafilm®
7) Small blenders

8) Whatman No. 1 filter paper
9) Knife
10) Ice bucket
11) Erlenmeyer flask
12) Beakers, 100 and 600 ml
13) Graduated cylinder, 50 ml
14) Paper towels
15) Forceps
16) Hot plate

8.4 Reagents and Materials

1) Small potatoes (held in refrigerator overnight)
2) Ice in an ice bucket
3) Buffer A: Sodium phosphate buffer, 0.1 M, pH 6.8, containing 0.1 M NaF (prepared by lab assistants)
4) Buffer B: Sodium phosphate buffer, 0.1 M, pH 6.8 (prepared by lab assistants)
5) Sulfite solution: 0.5% (wt/vol) $NaHSO_3$ in 0.1 M sodium phosphate buffer, pH 6.8
6) Dopa (DL-3,4-Dihydroxyphenylalanine). $4 \, mg \, ml^{-1}$ in buffer B. Freshly prepared by lab assistants. **Note**: It is difficult to dissolve dopa in pH 6.8 buffer. To facilitate solution, dissolve 400 mg of dopa in 10 ml of 0.1 N HCl, add 80 ml buffer B, adjust pH to 6.8 with 1.0 N KOH, dilute to 100 ml with buffer B. It may be necessary to heat slightly to effect solution in the 0.1 N HCl. Heat gently on a hot plate with constant stirring. Heat only until the dopa dissolves.

8.5 Procedures

Note: This procedure is an adaptation of a method described by Boyer [4].

8.5.1 Preparation of Crude Enzyme Extract

1) Peel a cooled potato and cut into small pieces.
2) Rapidly weigh about 10 g of potato and mix with 50 ml of ice cold buffer A.
3) Grind the mixture thoroughly in a blender until puréed.
4) Filter the mixture with Whatman No. 1 paper into an iced 125 ml Erlenmeyer flask and hold on ice until needed.

8.5.2 Enzyme Assay

1) Transfer the volumes of buffer B, sulfite solution, and dopa solution shown in Table 8.1 to cuvettes labeled 1 through 12 taking care to pipette as accurately as possible.
2) Set the wavelength on the spectrophotometer to 475 nm and select the enzyme assay method (if your spectrophotometer is equipped with an enzyme assay function). Zero the instrument against buffer B.
3) When everything is set, initiate the reaction by adding the enzyme extract to the cuvette and inverting to mix. Immediately place the cuvette in the spectrophotometer and begin recording data. Collect data for two minutes. Repeat with the next sample.

Table 8.1 Volumes of buffer, substrate, and enzyme extract for PPO assays.

Reagent	Cuvette number[a]											
	1	2	3	4	5	6	7	8	9	10	11	12
Buffer B[b]	2.5	2.4	2.35	2.3	2.2	2.1	2.0	1.9	0.8	0.7	0.6	—
Sulfite[c]	—	—	—	—	—	—	—	—	—	—	—	2.0
Dopa solution[d]	—	0.1	0.15	0.2	0.3	0.4	0.5	0.6	2.0	2.0	2.0	0.5
Enzyme extract	0.5	0.5	0.5	0.5	0.5	0.5	0.5	0.5	0.2	0.3	0.4	0.5

[a] All volumes are in ml. Total volume in each cuvette should equal 3.0 ml.
[b] 0.1 M phosphate buffer, pH 6.8.
[c] $NaHSO_3$ in buffer B, 0.5% (w/v).
[d] 4 mg dopa per ml buffer B, pH 6.8.

8.5.3 Data Treatment

1) Make a plot of absorbance versus time for each cuvette. Estimate the rate of reaction in each cuvette from the linear portion of the absorbance versus time curve.
2) Determine reaction velocities for cuvettes 2 through 8. Express reaction velocity as mmol IQ l^{-1} min^{-1}.
3) Construct a plot of velocity versus [S] for cuvettes 2 through 8.
4) Construct a Lineweaver-Burk plot from data for cuvettes 2 through 8. Determine K_M and V_{max} for your enzyme.
5) Determine the enzyme activities (rates) for cuvettes 9, 10, and 11. Plot activity vs. volume of enzyme extract used. Is your plot linear? If it is, is this what you expected? Why? If it is not, what might be a plausible explanation?
6) Compare the activity in cuvette 12 with that in cuvette 7. Are they different? Explain.

8.5.4 Required Notebook Entries

1) A plot of absorbance vs. time for each cuvette.
2) A table showing all data from the PPO assay. For each cuvette, you should record rate, velocity, 1/V, [S], and 1/[S]. Be sure to include units for each variable.
3) A plot of velocity vs. [S] for your PPO data.
4) A Lineweaver-Burk plot for your PPO data.
5) Calculations for V_{max} and K_M for PPO.
6) A plot of enzyme velocity vs. volume of enzyme extract for PPO data.

8.6 Problem Set

1   A crude extract of PPO was prepared by blending a small potato in buffer. The activity of the extract was assayed according to the protocol outlined above. Dopa (2.0 mmol l^{-1} in the final assay mixture) was used as the substrate. The following absorbance values (measured at 475 nm) were obtained:

Absorbance data from a PPO assay of a potato extract using dopa as the substrate									
Time (min)	0	0.25	0.5	0.75	1.0	1.25	1.5	1.75	2.0
A (475 nm)	0	0.1	0.2	0.3	0.39	0.47	0.54	0.60	0.65

a) Determine the initial rate of the reaction.
b) Calculate the velocity of the reaction.

2   A series of assays using varying substrate concentrations were conducted on the potato extract described in Problem 1 above. The following data were obtained:

Reaction velocities of the oxidation of dopa by PPO at varying substrate concentrations.			
[S] (mmol l^{-1})	1/[S] (l mmol^{-1})	V (mmol l^{-1} min^{-1})	1/V (l min mmol^{-1})
1.0		0.050	
1.5		0.069	
2.0		0.085	
2.5		0.099	
3.0		0.111	
3.5		0.125	
4.0		0.132	

a) Make a plot of velocity versus substrate concentration. Is it linear? Why or why not?
b) Construct a Lineweaver-Burk plot for the data. Determine the K_M and V_{max} for the extract.
c) Values for V_{max} often vary considerably between potato extracts. Can you think of an explanation? (**Hint**: the efficiency of enzyme extraction varies with blending times, temperature, and ratio of potato to buffer).

8.7   Study Questions

1   Is your Lineweaver-Burk plot linear? If not, what is your explanation?

2   What does your value for K_M indicate?

3   Is bisulfite an effective inhibitor?

4   Does enzyme concentration affect velocity? Would it affect K_M and V_{max} as they were determined in this experiment?

8.8   References

1 Singh, B., Suri, K., Shevkani, K. et al. (2018). Enzymatic browning of fruit and vegetables: a review. In: *Enzymes in Food Technology: Improvements and Innovations* (ed. M. Kuddus), 63–78. Singapore: Springer.

2 Parkin, K.L. (2017). Enzymes. In: *Fennema's Food Chemistry*, 5e (eds. S. Damodaran and K.L. Parkin), 357–465. Boca Raton: CRC Press, Taylor & Francis Group.

3 Sapers, G.M. (1993). Browning of foods: control by sulfites, antioxidants, and other means. *Food Technology* 47 (10): 75–84.

4 Boyer, R.F. (1977). Spectrophotometric assay of polyphenoloxidase activity. A special project in enzyme characterization. *Journal of Chemical Education* 54 (9): 585.

Answers to Problem Set

1 a) 0.4 absorbance units min^{-1}

  b) 0.08 mmol l^{-1} min^{-1}

2 b) $K_M = 5.22$ mmol l^{-1}; $V_{max} = 0.31$ mmol l^{-1} min^{-1}

9

Blanching Effectiveness

9.1 Learning Outcomes

After completing this exercise, students will be able to:

1) Test fruits and vegetables for blanching effectiveness.
2) Describe the biochemistry underlying the guaiacol test for blanching effectiveness.

9.2 Introduction

Fresh fruits and vegetables contain many active enzymes that cause postharvest deterioration in quality and nutritional value. This deterioration occurs even when the products are frozen. Thus, fruits and vegetables are usually blanched prior to freezing or canning in an attempt to inactivate these enzymes. Blanching may also kill some microorganisms that may be present in the fresh produce.

The thermal stabilities of enzymes vary widely as shown in Figure 9.1. Notice that it takes longer at a given temperature to inactivate peroxidase than, for example polyphenol oxidase. Therefore, blanching conditions need to be targeted to the most heat-resistant enzymes.

Peroxidases are ubiquitous in plant tissues and are among the most heat stable enzymes in plants [2]. Thus, they are good indicators for blanching adequacy since heat treatments sufficient to inactivate peroxidase also inactivate most other enzymes.

Peroxidases are oxidoreductases, i.e. they are members of the group of enzymes that catalyze oxidation–reduction reactions. As the name implies, one of the substrates of peroxidase is a peroxide. Several different peroxides may be present in plant tissues, including hydrogen peroxide and lipid hydroperoxides. Substrates capable of donating hydrogen atoms can reduce peroxides as shown in the following peroxidase-catalyzed reaction:

$$2 \text{ AH (a hydrogen atom donor)} + H_2O_2 \xrightarrow{\text{Peroxidase}} 2A\cdot + 2H_2O$$

In the above reaction, AH, a hydrogen donor, is oxidized by a peroxide, H_2O_2. Many peroxidases have low specificity, i.e. they catalyze the oxidation of many different hydrogen donors. Phenolic and other aromatic compounds are common substrates. Moreover, either a lipid hydroperoxide or hydrogen peroxide can function as the oxidizing agent. Peroxidases have several functions in plant tissues including removal of hydrogen peroxide, defense against pathogens and pests, and degradation of lignin [2].

Food Chemistry: A Laboratory Manual, Second Edition. Dennis D. Miller and C. K. Yeung.
© 2022 John Wiley & Sons, Inc. Published 2022 by John Wiley & Sons, Inc.
Companion website: www.wiley.com/go/Miller/foodchemistry2

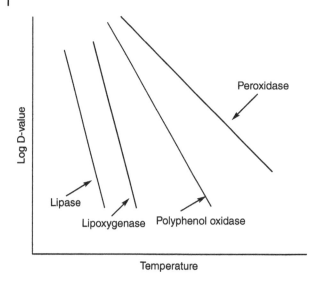

Figure 9.1 D-values for the thermal inactivation of selected enzymes at various temperatures. D-value is the time, at a given temperature, required to inactivate 90% of enzyme activity. *Source:* Adapted from [1].

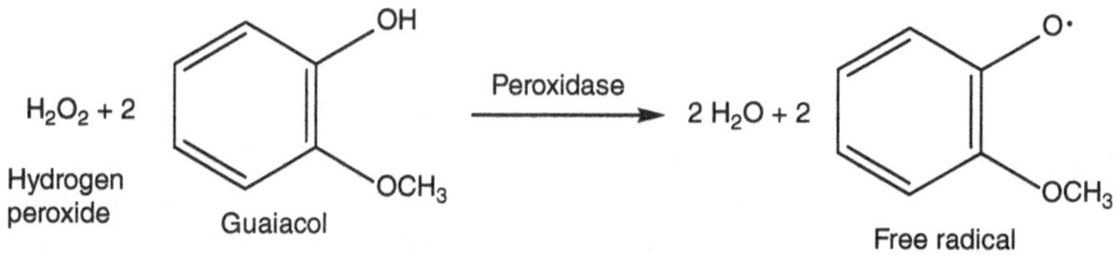

$H_2O_2 + 2$

Hydrogen
peroxide   Guaiacol   OH / OCH$_3$ $\xrightarrow{\text{Peroxidase}}$ $2\,H_2O + 2$ O· / OCH$_3$ Free radical

Figure 9.2 The oxidation of guaiacol by hydrogen peroxide to form a guaiacol-free radical. The reaction is catalyzed by peroxidase.

Guaiacol (o-methoxyphenol), in a reaction catalyzed by peroxidase, is oxidized by hydrogen peroxide to form free radicals (Figure 9.2).

The free radicals generated in the above reaction react to form compounds that have a reddish brown color and this is the basis for the peroxidase assay [2]. The structures of these compounds are in dispute, and it is likely that several different compounds form when guaiacol free radicals react with each other. Figure 9.3 shows one possible reaction sequence that yields 3,3′-dimethoxy-4,4′-biphenoq uinone. Given the multiple conjugated double bonds in the structure, we would expect it to be a colored compound.

The guaiacol peroxidase assay has been in use for decades and appears to be a reliable assay for blanching effectiveness even if there remains some doubt about the structure of the compound or compounds responsible for the red/brown color.

The above reactions provide a convenient, qualitative assay for blanching adequacy. The substrates, hydrogen peroxide and guaiacol, are mixed with macerated vegetables or fruits and incubated for a brief period. If a reddish brown color develops, then active peroxidase is present indicating inadequate blanching.

This blanching effectiveness assay does have its drawbacks. The primary objective of blanching is to inactivate enzymes that cause quality deterioration in the food. Peroxidases are not a major factor in quality deterioration, so heating to inactivate them may result in a more severe heat treatment than is necessary to prevent quality deterioration [1].

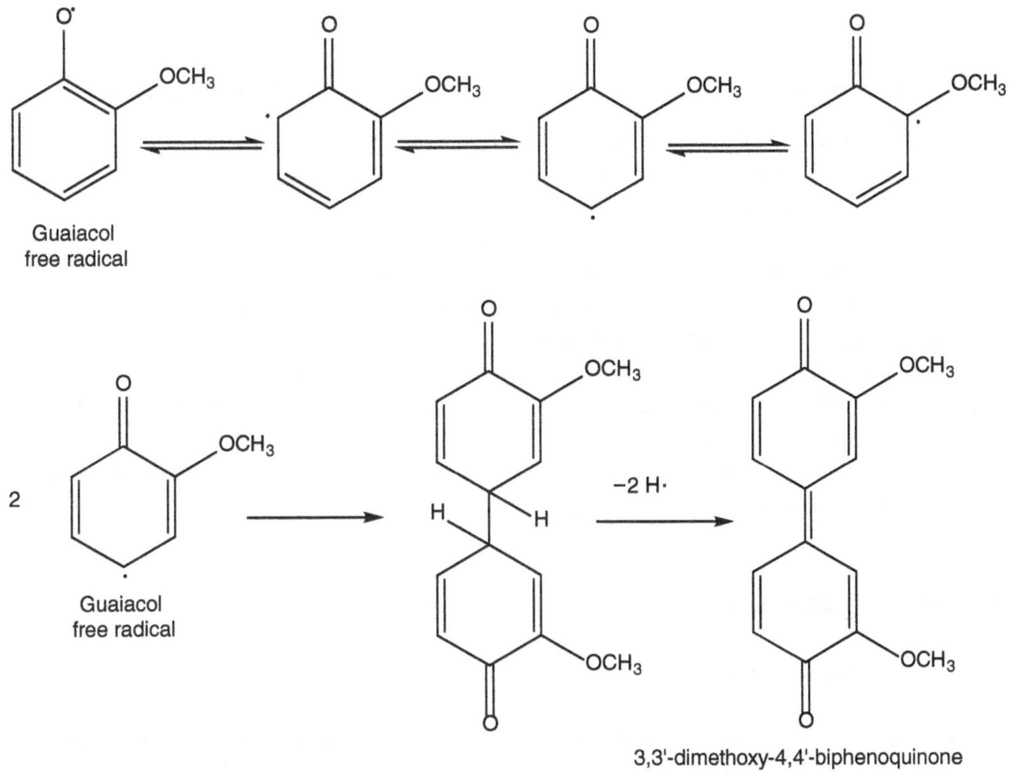

Figure 9.3 Possible reactions of guaiacol-free radicals that yield the red/brown color typical of a positive guaiacol test for blanching adequacy. *Source:* Adapted from Doerge et al. [3] and Hwang et al. [4].

9.3 Apparatus and Instruments

1) Beakers, 600 ml
2) Slotted spoon
3) Knife
4) Mortar and pestle
5) Graduated cylinder, 10 ml
6) Pipettes, 1 ml
7) Test tubes
8) Small beakers
9) Hot plate

9.4 Reagents and Materials

1) Fresh potatoes and apples
2) Guaiacol (1% v/v in 95% ethanol)
3) Hydrogen peroxide (0.5% v/v)
4) Sand

9.5 Procedures

The following procedure is adapted from the method described by Masure and Campbell [5].

1) Blanch a few pieces of potato and apple as follows: Bring 300 ml distilled water to boiling in a 600 ml beaker. Submerge pieces of each sample in the boiling water for two minutes. Remove and submerge in cold water to cool. Place on a paper towel.
2) Assay the potato and the apple before and after blanching as follows:
   a) Cut a piece of sample weighing about 5 g into small pieces.
   b) Transfer the sample to a mortar containing a small quantity of sand. Add about 5 ml distilled water and grind for two to three minutes.
   c) Add another 5 ml water, mix, and transfer contents to a small beaker.
   d) Add 1 ml of 1.0% guaiacol solution (in 95% ethanol) and 1 ml 0.5% hydrogen peroxide. Mix by gently swirling the beaker.
   e) Peroxidase activity is indicated by the development of a reddish color. If no color develops in 3.5 min, consider the product adequately blanched.

9.6 Study Questions

1 Why are some enzymes more stable to thermal treatment than others?

2 Name three enzymes that cause problems in frozen vegetables if they are not inactivated. Describe the reactions involved.

9.7 References

1 Belitz, H.-D., Grosch, W., and Schieberle, P. (2009). *Food Chemistry*, 4e. Berlin: Springer. 1070 p.
2 Parkin, K.L. (2017). Enzymes. In: *Fennema's Food Chemistry*, 5e (eds. S. Damodaran and K.L. Parkin), 357–465. Boca Raton: CRC Press, Taylor & Francis Group.
3 Doerge, D.R., Divi, R.L., and Churchwell, M.I. (1997). Identification of the colored guaiacol oxidation product produced by peroxidases. *Analytical Biochemistry* 250 (1): 10–17.
4 Hwang, S., Lee, C.-H., and Ahn, I.-S. (2008). Product identification of guaiacol oxidation catalyzed by manganese peroxidase. *Journal of Industrial and Engineering Chemistry* 14 (4): 487–492.
5 Masure, M.P. and Campbell, C. (1944). Rapid estimation of peroxidase in vegetable extracts - an index of blanching adequacy for frozen vegetables. *Fruit Products Journal and American Food Manufacturer* 23 (12): 369–374.

9.8 Suggested Reading

Burnette, F.S. (1977). Peroxidase and its relationship to food flavor and quality: a review. *Journal of Food Science* 42 (1): 1–6.
Xiao, H.-W., Pan, Z., Deng, L.-Z. et al. (2017). Recent developments and trends in thermal blanching – a comprehensive review. *Information Processing in Agriculture* 4 (2): 101–127.

10

Lipid Oxidation

10.1 Learning Outcomes

After completing this exercise, students will be able to

1) Describe the chemical processes involved in lipid oxidation in foods.
2) Use the thiobarbituric acid (TBA) test to analyze a food product for evidence of lipid oxidation.
3) Quantify the extent of lipid oxidation in a food from data generated in the TBA test and express as malondialdehyde (MDA) equivalents per gram of food (e.g. µg MDA g^{-1} meat).
4) Explain mechanisms responsible for warmed-over flavor development in meat.

10.2 Introduction

The oxidation of lipids in foods is a major cause of quality deterioration. Lipid oxidation (or peroxidation) causes off-flavors and odors, destroys sensitive vitamins, and may generate toxic compounds. It can affect most foods, even foods low in fat, if the conditions are right. The off-flavors and odors produced are frequently characterized as rancid, and hence the term oxidative rancidity.

Polyunsaturated fatty acids are especially susceptible to lipid peroxidation due to the double bonds in their hydrocarbon chains. This is because lipid peroxidation proceeds by a free radical mechanism, and unsaturated fatty acids are particularly susceptible to attack by free radicals.

10.2.1 The Chemistry of Lipid Oxidation

See Brady [1] or McClements and Decker [2] for in-depth discussions of lipid oxidation in foods and Dominguez et al. [3] for a review of lipid oxidation in meat and meat products.

To begin the process of lipid oxidation, a free radical must be generated. Hydroxyl, peroxyl, and hydroperoxyl radicals are among the most important initiator radicals in foods [4]. Hydroxyl radicals may be generated in foods in a number of ways including the Fenton reaction where ferrous iron reacts with hydrogen peroxide to form a hydroxyl radical:

$$Fe^{2+} + H_2O_2 \longrightarrow Fe^{3+} + OH^- + HO\cdot$$

Peroxyl, alkoxyl, and hydroxyl radicals are formed when hydroperoxides decompose (Figure 10.1).

Food Chemistry: A Laboratory Manual, Second Edition. Dennis D. Miller and C. K. Yeung.
© 2022 John Wiley & Sons, Inc. Published 2022 by John Wiley & Sons, Inc.
Companion website: www.wiley.com/go/Miller/foodchemistry2

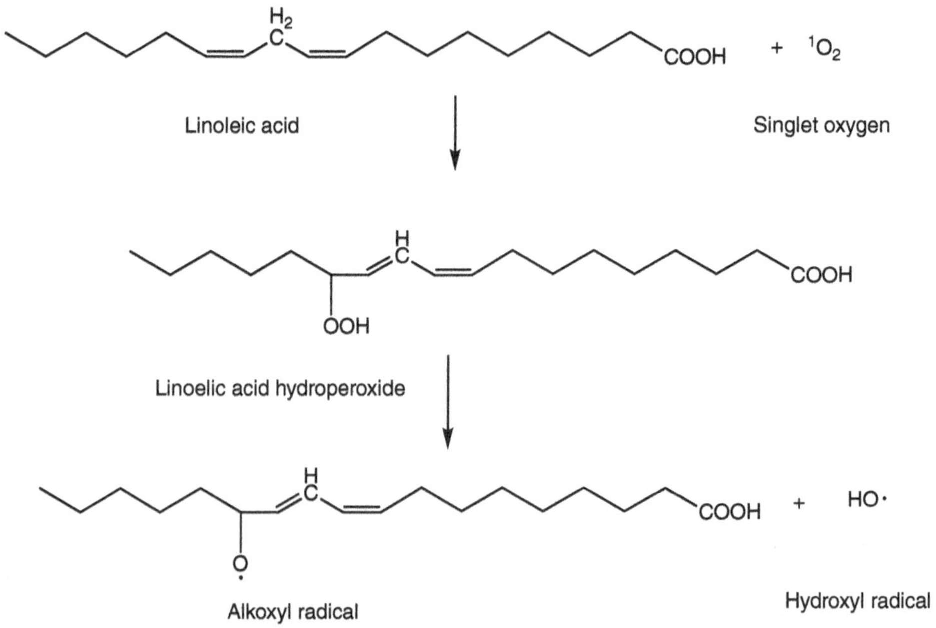

Figure 10.1 The reaction of linoleic acid with singlet oxygen to form a hydroperoxide and the subsequent decomposition of the hydroperoxide (initiation phase).

Hydroperoxides may form when singlet oxygen ($^1O_2$) attacks the double bond in an unsaturated fatty acid as shown in Figure 10.1. Atmospheric oxygen is called "ground state" oxygen because it is in the lowest energy state and is quite unreactive at ambient temperatures. It is classified as triplet oxygen ($^3O_2$) because its two outermost unpaired electrons have parallel spins. Triplet oxygen will not react with unsaturated fatty acids because of its spin state. However, triplet oxygen is a free radical (it has not one but two unpaired electrons in its outer shell) and therefore will readily react with other free radicals. Singlet oxygen may be formed in foods by photosensitization, which can occur when foods are exposed to light. Foods that contain light-absorbing pigments such as riboflavin and carotenoids are particularly susceptible to photosensitization.

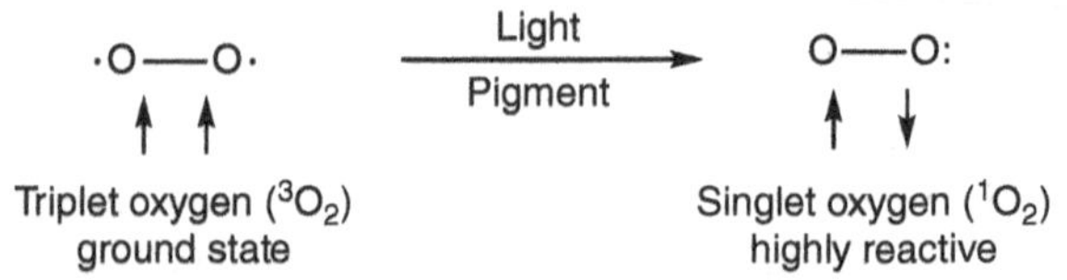

Once a lipid-free radical forms in a food, lipid oxidation can begin. Lipid oxidation is a complex chain reaction, i.e. products in the reaction are recycled so that the reaction is self-perpetuating. Using the example of linoleic acid, we see that hydrogen abstraction from carbon 11 yields a free radical (Figure 10.2). Polyunsaturated fatty acids like linoleic acid are particularly susceptible to free radical attack because the alkyl free radical that forms is stabilized by resonance. Triplet oxygen ($^3O_2$), unlike singlet oxygen, will not attack a double bond but reacts readily with free radicals to form peroxyl radicals. The peroxyl radical in turn can abstract a hydrogen from another linoleic acid molecule, thereby regenerating the alkyl free radical (R·).

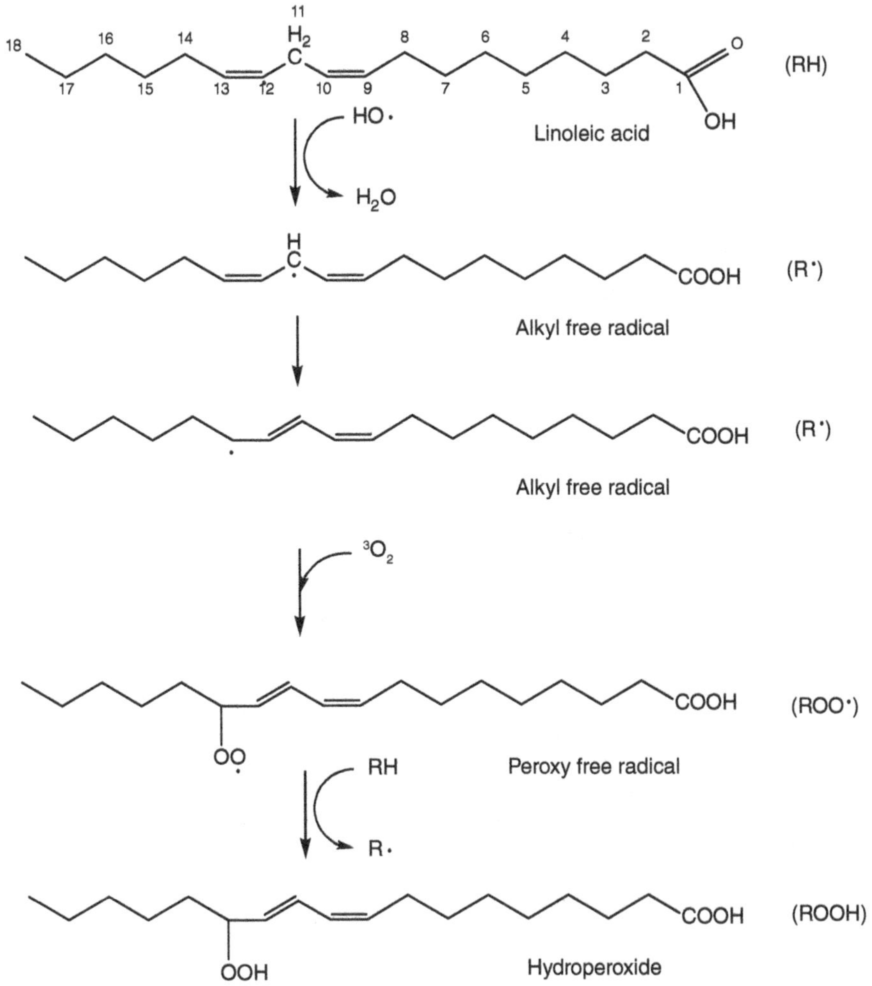

Figure 10.2 Reaction of linoleic acid with a hydroxyl radical and triplet oxygen to form a hydroperoxide (propagation phase).

Lipid oxidation is often characterized as having four stages: initiation, propagation, decomposition, and termination.

Lipid hydroperoxides are unstable molecules and are prone to undergo scissions of carbon–carbon bonds. Figure 10.3 shows scission products of a linoleic acid hydroperoxide. Some scission products, hexanal, for example, are volatile compounds and their concentrations in a headspace, determined using gas chromatography, can be used as indicators of the extent of lipid oxidation in foods.

Termination of the chain reaction occurs when two free radicals combine to form a single bond. A summary of the reactions believed to be involved in lipid oxidation is given in Figure 10.4.

Hydroperoxides are the initial products of lipid oxidation. As shown above, they decompose to aldehydes, ketones, alcohols, alkenes, and/or hydroxy acids. Such products frequently have unpleasant odors and flavors [1].

Off-flavors in meat and meat products arise in several ways. Frozen raw meat develops oxidative rancidity after several months of storage. Refrigerated cooked meats, on the other hand, often exhibit a "stale, painty, rancid" flavor and aroma upon reheating. This characteristic becomes

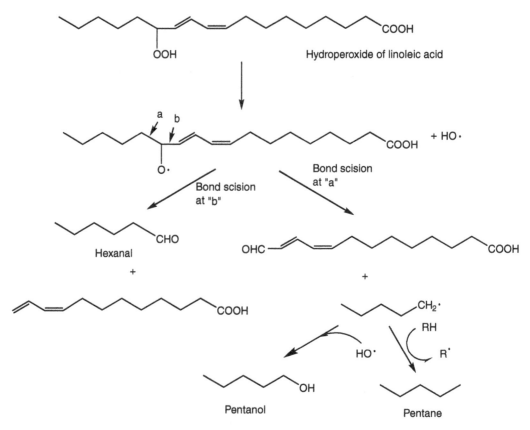

Figure 10.3 The decomposition of linoleic acid hydroperoxide (decomposition phase). Note that some scission products including hexanal and pentane are volatile compounds. Concentrations of these compounds in the headspace above a packaged food may be used to assess the extent of lipid oxidation.

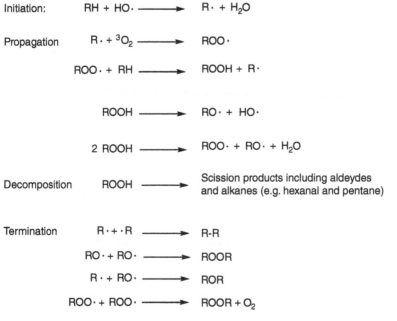

Initiation: $RH + HO\cdot \longrightarrow R\cdot + H_2O$

Propagation $\quad R\cdot + {}^3O_2 \longrightarrow ROO\cdot$

$\qquad\qquad ROO\cdot + RH \longrightarrow ROOH + R\cdot$

$\qquad\qquad\quad ROOH \longrightarrow RO\cdot + HO\cdot$

$\qquad\qquad\quad 2\,ROOH \longrightarrow ROO\cdot + RO\cdot + H_2O$

Decomposition $\quad ROOH \longrightarrow$ Scission products including aldeydes and alkanes (e.g. hexanal and pentane)

Termination $\quad R\cdot + \cdot R \longrightarrow R\text{-}R$

$\qquad\qquad RO\cdot + RO\cdot \longrightarrow ROOR$

$\qquad\qquad\quad R\cdot + RO\cdot \longrightarrow ROR$

$\qquad\qquad ROO\cdot + ROO\cdot \longrightarrow ROOR + O_2$

Figure 10.4 Reactions in lipid oxidation. RH is an unsaturated fatty acid. R⬤ is an alkyl free radical, ROO⬤ is an alkoxyl free radical, and ROOH is a lipid hydroperoxide.

evident after a few days storage at refrigeration temperatures and is known as **"warmed-over flavor (WOF)"** [5].

WOF has been associated with lipid oxidation. The greater the unsaturation of fatty acids in muscle tissue the greater the susceptibility to oxidative deterioration [6]. In general, pork and poultry contain more polyunsaturated lipids than meat from ruminants and, as a result, are more susceptible to lipid oxidation.

10.2.2 Control of Lipid Oxidation

It should be clear from the preceding sections that lipid peroxidation requires oxygen and free radicals. It is also known that transition metal ions such as iron and copper may initiate and accelerate lipid peroxidation by catalyzing the formation of new free radicals. Light may initiate peroxidation transforming triplet oxygen into singlet oxygen, thereby allowing the formation of lipid hydroperoxides, which can decompose to form free radicals. How then might lipid peroxidation in foods be controlled? Several strategies are available.

10.2.2.1 Elimination of Oxygen
Oxygen is required for lipid peroxidation. Therefore, elimination of oxygen should prevent peroxidation. This strategy is used in many food products, for example potato chips. It is accomplished by flushing away oxygen with nitrogen gas and sealing the food in packages that are impermeable to oxygen.

10.2.2.2 Scavenging of Free Radicals
Given that oxidation in foods proceeds by a free radical mechanism, it stands to reason that removal of free radicals would prevent or slow oxidation. In fact many of the most effective antioxidants in foods are free radical scavengers. To understand how free radical scavengers work, let's take phenol as an example. The hydrogen in the hydroxyl group of phenolic compounds is relatively easily abstracted. Thus, phenolic compounds will donate a hydrogen atom to free radicals converting them to non-radicals:

$$PhOH + R\bullet \longrightarrow PhO\bullet + RH$$

We still have a free radical, PhO•. However, phenolic free radicals are relatively stable and are incapable of abstracting a hydrogen from another unsaturated fatty acid. Thus, the chain is broken, and lipid oxidation is slowed. Phenolic radicals are stable because of resonance around the aromatic ring:

Although phenol itself is toxic and cannot be added to foods, many phenolic compounds (derivatives of phenol) are present naturally in foods. Vitamin E is an important antioxidant in blood and cell membranes. It is a phenolic compound:

alpha-Tochpherol (vitamin E)

Also, the widely used synthetic food additives BHT, BHA, TBHQ, and propyl gallate (PG) are phenolic compounds.

10.2.2.3 Chelation of Metal Ions

Since transition metal ions catalyze the formation of free radicals, we might expect that removal of metal ions would reduce oxidation. However, it is not possible or practical to remove them from foods. Iron and copper are both essential nutrients. Moreover, iron deficiency is a widespread nutritional problem. Therefore, even if it were possible to remove these metals from foods, it would be unwise to do so. In fact, iron is added to many foods in order to insure adequate intakes by individuals in the population. Therefore, chelating agents capable of complexing metals in foods are often added to reduce the prooxidant effects of metals. EDTA and citric acid are common chelating agents added to foods.

10.2.3 Measurement of Lipid Oxidation in Foods

Chemical and physical tests for measuring lipid oxidation are based on the determination of hydroperoxides or their decomposition products [7, 8]. While sensory evaluation is the ultimate test for consumer acceptance of a product [9], it is a time-consuming procedure involving trained panelists and results are often variable. Chemical methods are more objective although interferences and poor reproducibility are frequent problems. Most chemical methods are highly empirical. Different methods assess different stages of the oxidative process, so correlations with sensory data may differ depending on the chemical method chosen [10]. The use of advanced instrumental analysis such as GC-MS has been very promising in identifying volatile flavor compounds generated from lipid oxidation and linking them to sensory attributes [11, 12].

Chemical methods are based on the disappearance of reactants or the appearance of products [2, 10, 13].

10.2.3.1 Thiobarbituric Acid Test (TBA Test)

Malondialdehyde (MDA) ($CHOCH_2CHO$) is believed to be one of the products formed when hydroperoxides decompose. It reacts with TBA to form a pink-colored chromogen, which absorbs at 450, 530, and 538 nm. This is one of the more widely used tests and will be used for this laboratory experiment. The reaction between TBA and MDA is shown below:

TBA   MDA   TBA-MDA Adduct

    Since other aldehydes besides MDA may also react with TBA to give a pink color, the values obtained in the assay are frequently expressed as TBARS (ThioBarbituric Acid Reactive Substances) or as MDA equivalents (e.g. µg MDA g^{-1} meat). An important caveat is that the TBA reaction is somewhat nonspecific, so results need to be interpreted with caution. An advantage of the TBA test is that it is relatively simple and inexpensive. Also, when used properly, it provides useful information about lipid oxidation not only in foods but in blood samples from humans and experimental animals.

10.2.3.2 Peroxide Value
This test measures the concentration of peroxides. Because peroxides may decompose to other products prior to measurement, caution must be exercised in the interpretation of peroxide value data.

10.2.3.3 Conjugated Diene Methods
Polyunsaturated fatty acids have a 1,4-pentadiene structure (R-CH=CHCH$_2$-CH=CH-R). When one of the methylene hydrogens is abstracted to form a free radical, the double bonds can rearrange to form conjugated double bonds (R-CH-CH=CH-CH=CH-R) (see Figure 10.3). Conjugated dienes absorb ultraviolet light more strongly than nonconjugated dienes. Therefore, in the early stages of oxidation, ultraviolet absorbance increases. Subsequently, it decreases as hydroperoxides containing the conjugated double bonds decompose.

10.2.3.4 Oxygen Bomb Test
This test measures oxygen disappearance from a system.

10.2.3.5 Total and Volatile Carbonyl Compounds
This method measures compounds that are responsible for off-flavors and odors. Several analytical methods have been used. Interferences from hydroperoxides may be a problem.

10.2.3.6 Anisidine Value Test
The anisidine value test is particularly useful for abused oils with low peroxide values such as frying oils [7]. The test involves a condensation reaction between the conjugated dienals or alk-2-enals in the sample and p-anisidine reagent in iso-octane solution followed by spectrophotometric measurement at 350 nm.

10.3 Apparatus and Instruments

Note: It is a good idea to acid-wash all glassware to avoid contamination from trace metal ions, which can catalyze lipid oxidation during the assay giving variable results.

1) Screw-capped test tubes (30)
2) Glass pipettes & bulb, 1, 5, and 10 ml
3) Beakers, 250 ml, tall form
4) Graduated cylinder, 50 ml
5) Plastic funnels
6) Erlenmeyer flasks, 250 ml
7) Filter paper, Whatman #1, 18 cm diameter
8) Spectrophotometer and cuvettes
9) Hand-held blender
10) Vortex mixer

10.4 Reagents and Materials

1) Fresh, raw (red muscle) ground turkey
2) Cooked (red muscle) ground turkey, stored 48 hours at 4 °C
3) TBA reagent (0.02 M in dH_2O)
4) Extracting solution = 10% TCA in 0.2 M H_3PO_4, w/v
5) 1,1,3,3-Tetraethoxypropane (TEP) standard solution, 10^{-2} M
6) TEP, 10 μM
7) TEP, 25 μM
8) BHT, 0.2 mg ml^{-1}

10.5 Procedures: Lipid Oxidation in Turkey Meat

Note: The following procedure applies to the analyses of both raw (fresh) and cooked (stored) ground turkey samples. Two additional samples (1 raw and 1 cooked) will be spiked with TEP in order to assess recovery of TBARS.

1) Weigh 5 g of raw turkey into each of three 400 ml glass beakers. Repeat with 5 g samples of cooked turkey. Weights need not be exactly 5 g but record exact weights for all six turkey samples.
2) Add 1 ml BHT (0.2 mg ml^{-1}) to each of your 6 pre-weighed turkey samples. To 1 raw and 1 cooked sample, add 12 ml of 10 μM TEP.
3) Add 33.5 ml TCA/H_3PO_4 to each of the TEP-spiked samples, bringing the total volume to 50 ml (assume moisture content of all turkey samples is 70% or 3.5 ml H_2O in 5 g). Add 45.5 ml TCA/H_3PO_4 to each of the four remaining beakers.
4) Blend the contents of each beaker for 30 seconds on HIGH using a hand-held blender until thoroughly blended. CAUTION! Start blending at a low speed and gradually increase speed to prevent splashing. Shake off liquid and any meat pieces before blending next sample. Rinse blender in a large beaker of dH_2O between samples.

5) Filter each blended meat mixture through Whatman #1 filter paper, collecting filtrate in pre-labeled 250 ml Erlenmeyer flasks. Proceed to Step 6 while samples are filtering.

6) Prepare standards for a *standard curve* as follows:
   a) Prepare a table like the one in Section 10.6 on the following page to keep the concentrations in your standards straight.
   b) Pipette aliquots of 0, 0.5, 1.0, 1.5, and 2.0 ml of 25 µM TEP into screw-cap test tubes, in duplicate.
   c) Bring total volume of each tube to 5 ml with TCA/H_3PO_4 solution.
   d) Add 5 ml of TBA reagent to each tube, cap, mix, and set aside.

7) Remove funnels from flasks in Step 5, swirl each flask to mix. Pipette two 5 ml aliquots of each filtrate into 2 separate screw-cap test tubes.
   a) To one aliquot, add 5 ml TBA Reagent. (This is your *test sample*).
   b) To the second aliquot, add 5 ml dH_2O. (This is your *sample blank*).

8) Prepare a reagent blank by pipetting 5 ml of the extracting solution and 5 ml of the TBA reagent into a test tube.

9) Cap tubes, vortex to mix, and place *all tubes* in a dark cabinet for 15–20 hours.

10) Read absorbances at 530 nm against the reagent blank. Subtract sample blank absorbances from sample readings only. It may be necessary to dilute cooked turkey samples so that absorbances lie in the linear range of the standard curve.

11) Construct a standard curve by plotting Absorbance vs. concentration of MDA (as nmol MDA ml^{-1}).

12) Compute MDA recovery (express as a percentage).

13) Determine the concentration of TBARS in the meat samples (as µg MDA per g of meat). Assume recovery of TBARS is the same as for MDA and make the appropriate correction.

14) Calculate means and standard deviations for TBARS in the meat samples from class data.

10.6 Problem Set: Calculation of TBARS

TBARS are often expressed as the "TBA Number." The TBA number for meat samples is equal to µg MDA g^{-1} tissue. The TBA number for samples is derived from absorbance readings and a standard curve. The standard curve is constructed from absorbance data from a series of standards (see Step 6 in Section 10.5).

Since MDA is a volatile substance and therefore difficult to handle, a nonvolatile precursor is used. We will use 1,1,3,3-tetraethoxypropane (TEP) as the precursor. TEP is a liquid at room temperature with a boiling point of 220 °C. TEP decomposes to MDA as follows:

Given the following data shown in Tables 10.1 and 10.2, calculate TBA numbers for raw and cooked turkey meat. Assume that the protocol outlined above was followed. **Hint**: For the turkey samples, first calculate the concentration of MDA in the TCA/H_3PO_4 extract. Then determine the

Table 10.1 Volumes, concentrations, and absorbance values for standards in the TBARS assay. The TEP standards were added to tubes, diluted to a volume of 5.0 ml with TCA/H_3PO_4 solution, mixed with 5.0 ml TBA reagent, and held for 15–20 hours before reading absorbances.

TEP (ml)	MDA (nmol tube^{-1})	[MDA] (nmol ml^{-1})^a	Absorbance (530 nm)
0.0	0	0	0.00
0.5	12.5	2.5	0.19
1.0	25.0	5.0	0.41
1.5	37.5	7.5	0.59
2.0	50.0	10.0	0.81

^{a} nmol ml^{-1} in the diluted standards before addition of TBA reagent.

Table 10.2 Data for TBARS assay on raw and cooked turkey samples.

Sample	Sample Wt (g)	Absorbance at 530 nm			[MDA] in extract		[MDA] in meat	
		Blank (no TBA)	Test (+ TBA)	Test – blank	nmol ml^{-1}	nmol 50 ml^{-1}	µg in sample	µg g^{-1}
Raw Turkey	5.1	0.01	0.1	0.09				
Raw + TEP	5	0.02	0.29	0.27				
Cooked Turkey	4.9	0.01	0.50	0.49				
Cooked + TEP	5	0.01	0.65	0.64				

total amount of MDA in the extract. This is equal to the total amount of MDA in the sample of meat that was extracted. Express the TBA number as µg MDA g^{-1} meat. Don't forget to correct your TBA numbers for recovery. Recall that 1 µmole = 1,000 nmole.

10.7 Study Questions

1 Why was BHT added prior to blending?

2 What was the purpose of TCA and H_3PO_4 in the extracting solution?

3 Why is the x-axis on the standard curve labeled as MDA ml^{-1} rather than TEP?

4 Which samples had the highest TBA numbers, raw or cooked? Explain why one was higher than the other.

10.8 References

1 Brady, J.W. (2013). *Introductory Food Chemistry*. Ithaca: Comstock Publishing Associates. 638 p.

2 McClements, D.J. and Decker, E.A. (2017). Lipids. In: *Fennema's Food Chemistry*, 5e (eds. S. Damodaran and K.L. Parkin), 171–233. Boca Raton: CRC Press, Taylor & Francis Group.

3 Domínguez, R., Pateiro, M., Gagaoua, M. et al. (2019). A comprehensive review on lipid oxidation in meat and meat products. *Antioxidants* 8 (10): 429.

4 Laguerre, M., Lecomte, J., and Villeneuve, P. (2007). Evaluation of the ability of antioxidants to counteract lipid oxidation: existing methods, new trends and challenges. *Progress in Lipid Research* 46 (5): 244–282.

5 St. Angelo, A.J. (1987). Preface. In: *Warmed-Over Flavor of Meat [Internet]* (eds. A.J. St. Angelo and M.E. Bailey), vii–viii. Orlando, FL: Academic Press http://qut.eblib.com.au/patron/FullRecord. aspx?p=1180688.

6 Cross, H.R. and Overby, A.J. (eds.) (1988). *Meat Science, Milk Science, and Technology*. Amsterdam; New York: Elsevier Science Publishers. 458 p. (World animal science).

7 Robards, K., Kerr, A.F., and Patsalides, E. (1988). Rancidity and its measurement in edible oils and snack foods. A review. *Analyst* 113 (2): 213–224.

8 Moon, J.-K. and Shibamoto, T. (2009). Antioxidant assays for plant and food components. *Journal of Agricultural and Food Chemistry* 57 (5): 1655–1666.

9 Arroyo, C., Eslami, S., Brunton, N.P. et al. (2015). An assessment of the impact of pulsed electric fields processing factors on oxidation, color, texture, and sensory attributes of turkey breast meat. *Poultry Science* 94 (5): 1088–1095.

10 Gray, J.I. (1978). Measurement of lipid oxidation: a review. *Journal of the American Oil Chemists' Society* 55 (6): 539–546.

11 Tikk, K., Haugen, J.-E., Andersen, H.J., and Aaslyng, M.D. (2008). Monitoring of warmed-over flavour in pork using the electronic nose – correlation to sensory attributes and secondary lipid oxidation products. *Meat Science* 80 (4): 1254–1263.

12 Chen, C., Husny, J., and Rabe, S. (2018). Predicting fishiness off-flavour and identifying compounds of lipid oxidation in dairy powders by SPME-GC/MS and machine learning. *International Dairy Journal* 77: 19–28.

13 Melton, S.L. (1983). Methodology for following lipid oxidation in muscle foods. *Food Technology (USA)* 37 (7): 105–116.

10.9 Suggested Reading

Byrne, D.V., O'Sullivan, M.G., Dijksterhuis, G.B. et al. (2001). Sensory panel consistency during development of a vocabulary for warmed-over flavour. *Food Quality and Preference* 12 (3): 171–187.

Jackson, V. and Penumetcha, M. (2019). Dietary oxidised lipids, health consequences and novel food technologies that thwart food lipid oxidation: an update. *International Journal of Food Science & Technology* 54 (6): 1981–1988.

Wilson, B.R., Pearson, A.M., and Shorland, F.B. (1976). Effect of total lipids and phospholipids on warmed-over flavor in red and white muscle from several species as measured by thiobarbituric acid analysis. *Journal of Agricultural and Food Chemistry* 24 (1): 7–11.

Zhang, Y., Holman, B.W.B., Ponnampalam, E.N. et al. (2019). Understanding beef flavour and overall liking traits using two different methods for determination of thiobarbituric acid reactive substance (TBARS). *Meat Science* 149: 114–119.

Answers to Problem Set

Equation for standard curve: $y = 0.081x - 0.004$, $R^2 = 1.00$.

Recovery $= 93.5$ and 72% for raw and cooked, respectively.

TBA no. (uncorrected) $= 0.82$ and $4.48\,\mu g\,g^{-1}$ for raw and cooked turkey, respectively.

TBA no. (corrected) $= 0.88$ and $6.22\,\mu g\,g^{-1}$ for raw and cooked turkey, respectively.

11

Ascorbic Acid: Stability and Leachability

11.1 Learning Outcomes

After completing this exercise, students will be able to:

1) Describe functions of ascorbic acid as a food additive.
2) Measure the concentration of ascorbic acid in a food product using a redox titrimetric procedure.
3) Determine the stability of ascorbic acid under various conditions.
4) Determine leaching losses in a common cooking method.

11.2 Introduction

L-Ascorbic acid (vitamin C) is an essential nutrient for humans. Low intakes cause a nutrient deficiency disease known as scurvy. The Recommended Dietary Allowance (RDA) for ascorbic acid is 90 mg day^{-1} for adult males and 75 mg day^{-1} for adult females [1]. About 10 mg day^{-1} will prevent scurvy, but higher intakes have health benefits beyond the prevention of scurvy [2].

Ascorbic acid is present naturally in many fruits and vegetables. Synthetic ascorbic acid is classified by FDA as a GRAS (generally recognized as safe) food additive. It is added to a wide variety of foods for both nutritional and technical reasons (Table 11.1).

11.2.1 Chemistry

Naturally occurring ascorbic acid is designated as L-ascorbic acid. As shown in Figure 11.1, there are four stereoisomers of ascorbic acid. D-ascorbic acid has 1/10 the vitamin activity of L-ascorbic acid and the isoascorbic acids have only 1/20 the activity. The FDA requires that the isoascorbic acids be listed by their common name, "erythorbic acid," to avoid misunderstanding by the consumer [3].

Ascorbic acid is quite acidic even though it has no free carboxyl groups like most other organic acids found in foods. The hydrogen on the hydroxyl on carbon 3 is the acidic hydrogen as shown in Figure 11.2.

Another interesting aspect of ascorbic acid is that it is an unusually stable lactone. Recall that a lactone is an ester where both the carboxyl and hydroxyl groups that form the ester are on the same

Food Chemistry: A Laboratory Manual, Second Edition. Dennis D. Miller and C. K. Yeung.
© 2022 John Wiley & Sons, Inc. Published 2022 by John Wiley & Sons, Inc.
Companion website: www.wiley.com/go/Miller/foodchemistry2

Table 11.1 Functions of naturally occurring and added ascorbic acid (or its isomers and derivatives) in foods.

Function	Food application
Essential nutrient	Fruit juices, fruit drinks, breakfast cereals
Antioxidant (controls oxidative rancidity and/or prevents enzymatic browning)	Apples, peaches, apricots, potatoes, cauliflower, mushrooms, olives, nuts, tomatoes, peanut butter, potato chips, soft drinks, fruit drinks
Antioxidant (added as ascorbyl palmitate)	Vegetable oils (acts synergistically with phenolic antioxidants such as tocopherols, BHA, and BHT to prevent oxidative rancidity)
Inhibition of can corrosion	Soft drinks
Protection of taste, flavor, clarity	Wines
Prevention of black spots	Shrimp
Inhibition of nitrosamine formation	Bacon
Color development in cured meats	Bacon, ham, hot dogs, sausages
Dough improver	Wheat flours for baked goods
Color stability in packaged meat	Fresh pork (Stabilizes the reddish color. Not permitted in other meats because it may mislead the consumer)

Source: Adapted from [3].

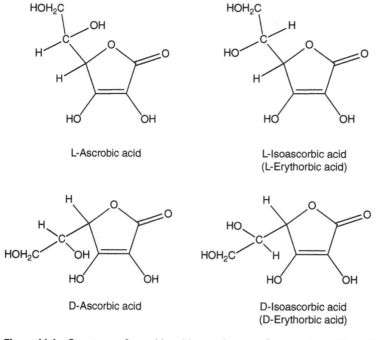

Figure 11.1 Structures of ascorbic acid stereoisomers. *Source:* Adapted from [4].

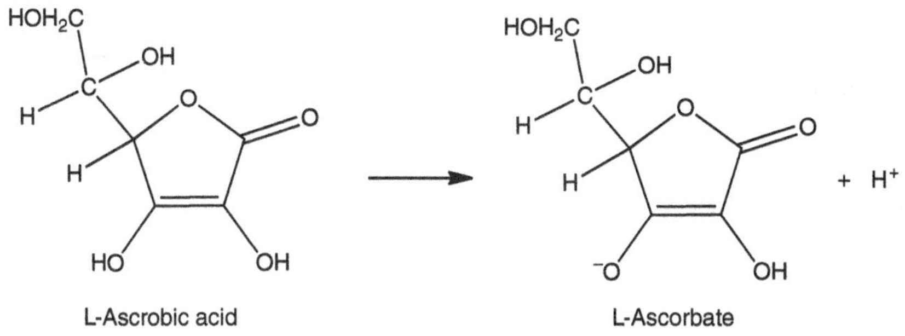

Figure 11.2   The dissociation of ascorbic acid to form ascorbate ion and a proton. The pKa of the C-3 hydroxyl group is 4.04.

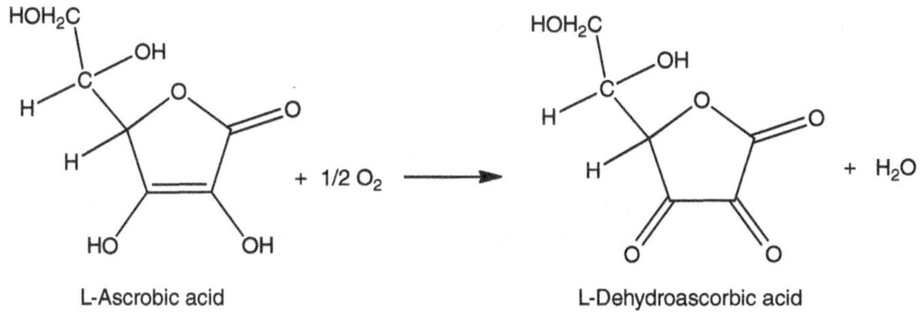

Figure 11.3   The reaction of ascorbic acid with molecular oxygen.

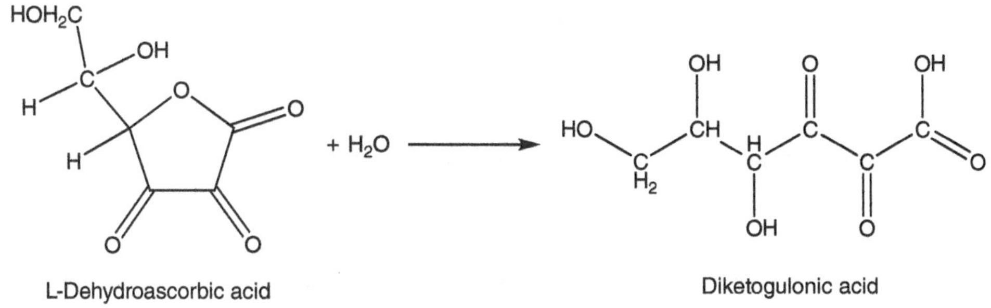

Figure 11.4   The hydrolysis of dehydroascorbic acid to diketogulonic acid.

molecule. This stability is lost when ascorbic acid is oxidized to dehydroascorbic acid (Figure 11.3). Dehydroascorbic acid readily hydrolyses to diketogulonic acid (Figure 11.4). Other properties of ascorbic acid are listed in Table 11.2.

11.2.2   Functions of Ascorbic Acid in Foods [3, 5, 6]

11.2.2.1   Oxygen Scavenger

One mechanism for the antioxidant activity of ascorbic acid in foods is oxygen scavenging. When foods are bottled or canned, they often contain residual molecular oxygen, which could react with

Table 11.2 Chemical and physical properties of ascorbic acid.

Strong reducing agent, effective antioxidant

Molecular weight = 176

Water solubility: 33% wt/v at 25 °C

Essential nutrient – prevents scurvy; RDA = 90 mg for men, 75 mg for women

Weak acid: $pk_{a1} = 4.2$; $pk_{a2} = 11.8$

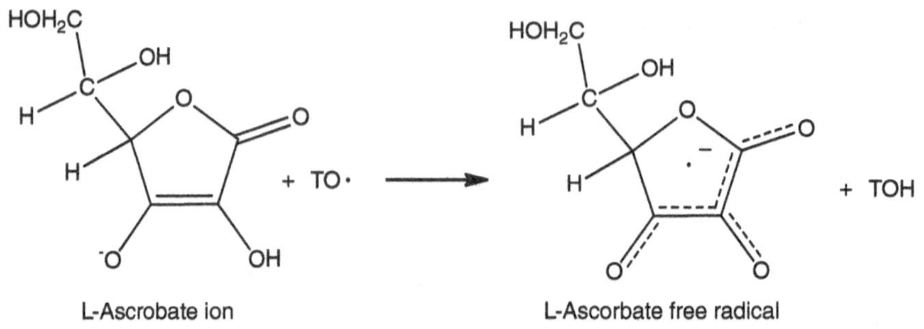

L-Ascrobate ion   L-Ascorbate free radical

Figure 11.5 The mechanism for free radical scavenging by ascorbate. TO·is tocopherol radical, TOH is tocopherol. The ascorbate radical is stabilized by resonance as shown. *Source:* Adapted from [7].

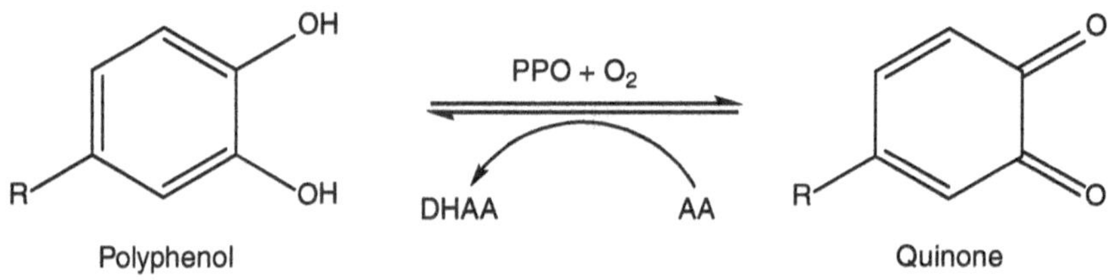

Polyphenol   Quinone

Figure 11.6 Inhibition of enzymatic browning by ascorbic acid. PPO = polyphenol oxidase, AA = ascorbic acid, DHAA = dehydroascorbic acid.

various food molecules to cause rancidity, loss of color, etc. Added ascorbic acid can remove or scavenge this oxygen as shown in Figure 11.3.

11.2.2.2 Free Radical Scavenger

Ascorbate ion scavenges free radicals by donating hydrogen atoms to other free radicals such as a tocopherol radical as shown in Figure 11.5. Thus, ascorbate acts synergistically with vitamin E and other phenolic antioxidants such as BHA and BHT. It is frequently added to foods along with phenolic antioxidants.

11.2.2.3 Control of Enzymatic Browning

Ascorbic acid is widely used to control enzymatic browning. This is accomplished by reducing quinones formed by the polyphenol oxidase-catalyzed oxidation of polyphenolics in foods as shown in Figure 11.6.

11.2.2.4 Dough Improver

Ascorbic acid is widely used as a dough improver in bread baking. Dough improvers are added to flour to strengthen the gluten and increase bread volume. Presumably, dough improvers function by preventing the reduction of disulfide bridges that cross link gluten proteins. Disulfide bonds may be broken by reactions known as disulfide interchanges where low-molecular-weight-free thiol groups react with disulfide bonds in the gluten proteins:

$$\text{Protein-S-S-Protein} + 2\,\text{R-SH} \longrightarrow 2\,\text{Protein-S-H} + \text{R-S-S-R}$$

This reaction reduces protein–protein interactions and thereby softens or weakens the dough. It follows that reducing the concentration of R-SH will slow the above reaction, and this will prevent or reduce dough softening. Glutathione, a tripeptide containing the amino acid cysteine, is one source of free thiol groups in bread flour, so any reaction that might reduce the concentration of glutathione may be expected to improve dough. Since thiol groups are easily oxidized, we might expect that oxidizing agents may improve doughs, and this, in fact, is the case. Two oxidants commonly used as flour additives are $KBrO_3$ and KIO_3. Ascorbic acid is also used, but it is an antioxidant, not an oxidant. How then can it be a dough improver? The explanation lies in its conversion to dehydroascorbic acid by ascorbic acid oxidase which is naturally present in wheat flour. Dehydroascorbic acid is the oxidized form of ascorbic acid and therefore is an oxidizing agent, capable of oxidizing other compounds. Figure 11.7 summarizes the reactions that are presumably involved in the dough improving action of ascorbic acid.

11.2.3 Stability of Ascorbic Acid

The effectiveness of ascorbic acid as a food additive depends on its oxidation to dehydroascorbic acid. Dehydroascorbic acid still has vitamin C activity and is the active agent in the dough improving function of ascorbic acid. Dehydroascorbic acid is a lactone but, unlike ascorbic acid, is readily hydrolyzed (Figure 11.4).

Diketogulonic acid has no vitamin activity and no antioxidant activity. Thus, oxidation of ascorbic acid effectively destroys ascorbic acid. Since it is so readily oxidized, we might expect ascorbic acid to be an unstable vitamin. It is in fact quite unstable and is frequently called the most unstable vitamin.

Ascorbic acid may degrade via a number of different mechanisms. Anaerobic as well as aerobic pathways have been identified but when oxygen is present, oxidative degradation predominates. Factors, which may influence the rate of ascorbic acid degradation, include temperature, salt and sugar concentration, pH, oxygen concentration, metal catalysts (mainly iron and copper ions), and enzymes.

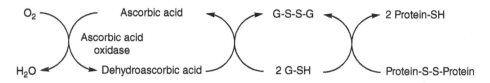

Figure 11.7 Reactions showing how ascorbic acid functions as a dough improver. Ascorbic acid is oxidized to dehydroascorbic acid in a reaction catalyzed by ascorbic acid oxidase. Dehydroascobic acid reacts with glutathione (G-SH) to form reduced glutathione (G-S-S-G). This reduces the concentration of G-SH, thereby reducing the breakage of disulfide bridges in the gluten proteins as shown at the far right in the figure. *Source:* Redrawn from [8].

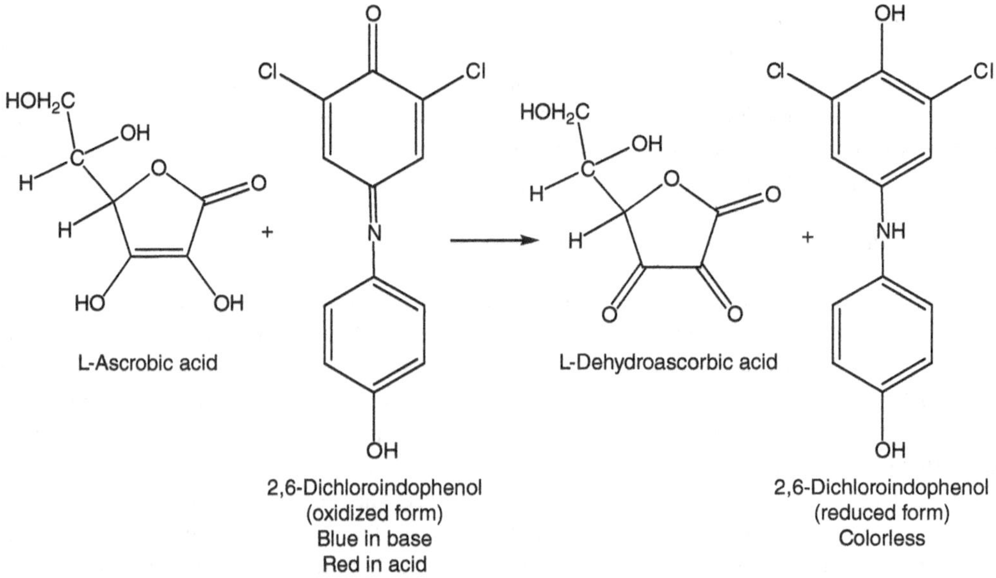

Figure 11.8 Reaction describing the redox titration of ascorbic acid with 2,6-dichloroindophenol.

11.2.4 Rationale for the Experiment

In this experiment, we will use a model system (an aqueous solution of ascorbic acid) to study the stability of ascorbic acid under different conditions.

Chemical degradation of ascorbic acid is only one route of loss from foods. Ascorbic acid is very soluble in water ($0.3\,g\,ml^{-1}$) and, as a result, leaching losses may be substantial. We will study the loss of ascorbic acid in cooked cabbage.

There are several assays available for the determination of ascorbic acid. Many of them are based on the strong reducing capacity of the vitamin. In this experiment, ascorbic acid will be determined by titrating with 2,6-dichloroindophenol (DIP) [9]. The reaction of DIP with L-ascorbic acid is shown in Figure 11.8.

When titrating an acid solution of ascorbic acid against blue DIP, the blue reagent will turn colorless when it is reduced by the ascorbic acid. When all the ascorbic acid has been oxidized, excess DIP will turn the solution pink, thereby showing the endpoint.

11.3 Apparatus and Instruments

1) Test tubes
2) Burettes
3) Beakers, 250 ml
4) Erlenmeyer flasks, 50 ml
5) Graduated cylinders, 10 and 50 ml
6) Pipettes, 1, 5, and 10 ml
7) Stirring plate
8) Water bath, 95 °C
9) pH meter

10) pH paper
11) Whatman No. 1 filter paper
12) Funnel
13) Sand
14) Mortar and pestle

11.4 Reagents and Materials

1) Ascorbic acid solution, $0.5 \, mg \, ml^{-1}$
2) Ascorbic acid solution, $10.0 \, mg \, ml^{-1}$
3) Oxalic acid solution, $0.25 \, M$
4) 2,6-dichloroindophenol (DIP): 250 mg 2,6-dichloroindophenol (Na salt) plus 210 mg $NaHCO_3$ per liter. **Note**: Dissolve the $NaHCO_3$ first, then add the dye. Vigorous shaking may be required to dissolve the dye. Store in amber bottle.
5) Cupric sulfate solution, $0.1 \, M$ (prepare using $CuSO_4 \cdot 5H_2O$ and distilled water)
6) Glycine buffer, $0.1 \, M$, pH 2
7) Phosphate buffer, $0.1 \, M$, pH 8
8) HCl, $1.0 \, N$
9) Fresh green cabbage

11.5 Procedures

11.5.1 Ascorbic Acid Standard Curve

1) Transfer approximately 9 ml of oxalic acid solution and 1 ml of $1.0 \, N$ HCl to each of four 50 ml Erlenmeyer flasks. Add 0.5, 1.0, 1.5, and 2.0 ml of ascorbic acid solution ($0.5 \, mg \, ml^{-1}$) to each flask, respectively
2) Titrate each flask rapidly with dye solution until a light but distinct rose-pink color persists for at least five seconds.
3) Plot volume of dye versus mg ascorbic acid. The plot should be linear. If not, repeat Steps 1 and 2. **Note**: Make plot *before* proceeding to Section 11.5.2.

11.5.2 Effect of pH on Ascorbic Acid Stability

1) Prepare in advance two ascorbic acid solutions, A and B:
   Solution A: $1.0 \, mg \, ml^{-1}$ ascorbic acid in $0.1 \, M$ glycine buffer, pH 2
   Solution B: $1.0 \, mg \, ml^{-1}$ ascorbic acid in $0.1 \, M$ phosphate buffer, pH 8

   Stir the solutions gently on a magnetic stir plate for 24 hours at room temperature. (**Note**: this step will be done by the teaching assistants.)
2) Using the procedure described in Section 11.5.1, determine the ascorbic acid concentration of Solutions A and B. The amount of ascorbic acid actually titrated should fall within the range of ascorbic acid used to construct the standard curve. Since the concentrations of ascorbic acid in Solutions A and B are unknown, you may need to do more than one titration before determining an appropriate amount to add. Try 1.0 ml of A and 4.0 ml of B to start.

11.5.3   Effects of Temperature, pH, and Cu^{2+} on the Stability of Ascorbic Acid

1) Prepare* 10 ml (in duplicate, in test tubes) of each of the following solutions:
   a) Ascorbic acid in glycine buffer (pH 2)
   b) Ascorbic acid in phosphate buffer (pH 8)
   c) Ascorbic acid in glycine buffer (pH 2) + CuSO$_4$
   d) Ascorbic acid in phosphate buffer (pH 8) + CuSO$_4$
   *To prepare the above solutions, add 1 ml ascorbic acid (10 mg ml^{-1}), 0.5 ml of 0.1 M CuSO$_4$ (where indicated) and sufficient buffer (pH 2 or pH 8) to bring the total volume to 10 ml. Mix well.
2) Measure and record the pH (use pH paper or pH meter) of each of your solutions. Cover or cap to reduce evaporation. Transfer tubes to a 95 °C water bath. Heat for 15 minutes, carefully remove tubes from the bath, and cool.
3) Using the procedure described in Section 11.5.1, determine (in duplicate) the ascorbic acid concentration in a 1 ml aliquot of solutions a–d.

11.5.4   Effect of Cooking on the Ascorbic Acid Content of Cabbage

1) Bring 100 ml distilled water to boiling in a 250 ml beaker. Add 5 g fresh cabbage and boil gently for 15 minutes. Remove the cabbage and place it on a paper towel to cool and drain. Measure volume of cooking water remaining in the beaker.
2) Transfer 5 g of fresh cabbage to a mortar. Add 10 ml of the oxalic acid solution. Add a pinch of sand. Macerate thoroughly (about three minutes). Add an additional 10 ml of the oxalic acid solution and again macerate. This is the *extract*. What is the total volume of your extract? (Assume that fresh and cooked cabbage are 92% water.)
3) Filter the macerated mixture through Whatman #1 filter paper. Collect the filtrate in a test tube.
4) Transfer 5 ml of the filtrate to a 50 ml Erlenmeyer flask. Add 9 ml oxalic acid solution plus 1 ml 1.0 N HCl. Titrate (in duplicate) to determine the ascorbic acid concentration.
5) Repeat Steps 2, 3, and 4 with the cooked cabbage. (Blot cooked cabbage with a paper towel, weigh, then macerate.)
6) Transfer 5 ml of the cooking water to a 50 ml Erlenmeyer flask. Add 9 ml of the oxalic acid solution plus 1 ml of 1.0 N HCl. Titrate (in duplicate) to determine the ascorbic acid concentration. Calculate the total amount of ascorbic acid in the cooking water.

11.6   Problem Set

1   Ascorbic acid standards were titrated according to the protocol described above. Given the following data (Table 11.3), plot the standard curve. Label the *y*-axis dichloroindophenol (DIP) (ml) and the *x*-axis ascorbic acid (mg).

2   Samples of cooked and raw cabbage were analyzed for ascorbic acid content according to the protocol described above. Data are shown in Table 11.4. Using the standard curve you constructed in Problem 1, calculate the concentration of ascorbic acid in the raw cabbage. Express your answer as mg AA/100 g cabbage. **Hint:** First determine the weight of ascorbic acid present in the 5.0 ml of filtrate that was titrated. Then calculate the total volume of the extract (the sum of the volumes of oxalic acid solution added to the cabbage plus the volume of water in

Table 11.3 Data for ascorbic acid (AA) standard curve.

Vol. AA (0.5 mg ml⁻¹) titrated (ml)	Wt. AA titrated (mg)	Vol DIP at endpoint (ml)
0.5	0.25	1.9
1.0	0.5	4.1
1.5	0.75	5.8
2.0	1.0	8.2

Table 11.4 Data from ascorbic acid titration experiment.

Sample	Wt or vol of Sample	Extract vol	Aliquot titrated (ml)	DIP at endpoint (ml)	AA in aliquot (mg)	AA in total sample (mg)	AA in cab. (mg/100 g)	AA as % of AA in raw
Raw cab.	5.0 g		5.0	5.5				100%
Cooked cab.	4.5 g		5.0	2.0				
Cook water	50 ml		10.0	1.5				

the cabbage). Finally, calculate the weight of ascorbic acid in the cabbage you extracted and then, using proportions, calculate the weight of ascorbic acid in 100 g of cabbage. Also calculate the amount of ascorbic acid that leached into the cooking water and the percentage of ascorbic acid in the raw cabbage that was in the cooked cabbage and the cooking water. Note that the weight of the cooked cabbage is less than 5 g. This is likely due to leaching of soluble components into the cooking water.

11.7 Study Questions

1 Based on your data, what was the route of greatest loss in the cooked cabbage, leaching or chemical degradation? Did you expect these results? Why or why not?

2 Is dehydroascorbic acid detected by this method? Why or why not? What effect might this have on the results that you obtained?

3 How do your values for ascorbic acid content of raw and cooked cabbage compare to handbook values? Give a possible reason for any discrepancies.

11.8 References

1 Otten, J.J., Hellwig, J.P., and Meyers, L.D. (eds.) (2006). *DRI, Dietary Reference Intakes: The Essential Guide to Nutrient Requirements*. Washington, DC: National Academies Press. 543 p.
2 Lykkesfeldt, J., Michels, A.J., and Frei, B. (2014). Vitamin C. *Advances in Nutrition* 5 (1): 16–18.

3 Newsome, R.L. (1987). Use of vitamins as additives in processed foods. *Food Technology* 41 (9): 163–168.

4 Gregory, J.F. III (2017). Vitamins. In: *Fennema's Food Chemistry*, 5e (eds. S. Damodaran and K.L. Parkin), 543–626. Boca Raton: CRC Press, Taylor & Francis Group.

5 Borenstein, B. (1987). The role of ascorbic acid in foods. *Food Technology* 41 (11): 98–99.

6 Chauhan, A.S., Ramteke, R.S., and Eipeson, W.E. (1998). Properties of ascorbic acid and its applications in food processing: a critical appraisal. *Journal of Food Science and Technology* 35 (5): 381–392.

7 Njus, D. and Kelley, P.M. (1991). Vitamins C and E donate single hydrogen atoms in vivo. *FEBS Letters* 284 (2): 147–151.

8 Belitz, H.-D., Grosch, W., and Schieberle, P. (2009). *Food Chemistry*, 4e. Berlin: Springer. 1070 p.

9 AOAC Official Method 967.21 (1968). *Ascorbic Acid in Vitamin Preparations and Juices: 2,6-dichloroindophenol Titrimetric Method*. AOAC International.

Answers to Problem Set

1 Equation for standard curve: $y = 8.24x - 0.15$; $R^2 = 0.996$.

2 Concentration of ascorbic acid in raw cabbage = 67.4 mg AA/100 g cabbage;
Concentration of ascorbic acid in cooked cabbage = 28 mg AA/100 g cabbage;
Amount of ascorbic acid in cooking water = 1 mg AA;
Total ascorbic acid in cooked cabbage + cooking water = 2.26 mg AA;
Amount of ascorbic acid that was destroyed (oxidized) = 3.37 mg − 2.26 mg = 1.11 mg or 33%.

12

Hydrolytic Rancidity in Milk

12.1 Learning Outcomes

After completing this exercise, students will be able to:

1) Draw structures of triacylglycerols.
2) Define hydrolytic rancidity and distinguish it from oxidative rancidity.
3) Explain the chemistry involved in the development of hydrolytic rancidity in milk.
4) Describe the copper soap method for measuring the concentration of free fatty acids in milk.
5) Measure the extent of hydrolytic rancidity in milk samples that have been subjected to various forms of abuse.

12.2 Introduction

Hydrolytic rancidity in milk develops when lipases attack triacylglycerols in fat globules to produce free fatty acids (Figure 12.1) [1]. In freshly drawn milk, triacylglycerols are protected from lipases by the milk fat globule membrane. Conditions or processes that damage this membrane expose the triacylglycerols and may lead to lipolysis catalyzed by lipases naturally present in milk. Pasteurization inactivates these lipases, but bacteria growing in pasteurized milk may produce other active lipases. This explains why pasteurized milk stored in the refrigerator too long may become rancid. Temperature fluctuations in stored milk; agitation of raw milk at the farm, during transportation or in the processing plant; and contamination of pasteurized, homogenized milk with raw milk may lead to excessive lipolysis.

Milk fat has a unique fatty acid composition with significant amounts of short chain fatty acids (Table 12.1). It is these short-chain fatty acids that cause milk to become rancid upon hydrolysis. Unlike long-chain fatty acids like stearic and linoleic acids, short-chain fatty acids like butyric acid are volatile and therefore have a sensory impact when free in solution [3].

This laboratory exercise is designed to study the effects of these factors on rancidity development in raw and processed milk. The degree of rancidity development will be measured using a method developed by Shipe et al. [4]. The method measures free fatty acid concentration by extracting copper soaps of the free fatty acids into chloroform-heptane-methanol with subsequent spectrophotometric measurement of the copper concentration in the extract.

Food Chemistry: A Laboratory Manual, Second Edition. Dennis D. Miller and C. K. Yeung.
© 2022 John Wiley & Sons, Inc. Published 2022 by John Wiley & Sons, Inc.
Companion website: www.wiley.com/go/Miller/foodchemistry2

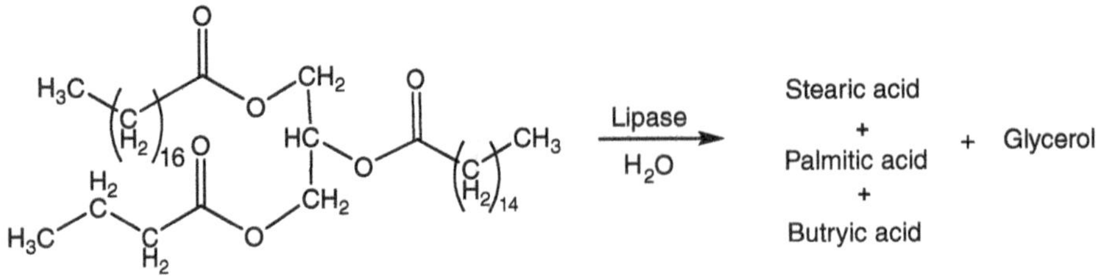

Figure 12.1 The lipase-catalyzed hydrolysis of a triacylglycerol (1-stearoyl-2-palmitoleoyl-3-butyroyl-glycerol) to yield free fatty acids plus glycerol.

Table 12.1 Fatty acid composition of bovine milk fat [2].

Fatty acid (FA)	Weight % of FA in milk fat
Butyric C4:0	4.8
Caproic C6:0	2.1
Caprylic C8:0	1.2
Capric C10:0	2.5
Lauric C12:0	3.2
Myristic C14:0	10.5
Myristoleic C14:1	1.1
Palmitic C16:0	29.9
Palmitelaidic C16:1 (trans)	0.3
Palmitoleic C16:1	2.0
Stearic C18:0	11.4
Elaidic C18:1 (trans)	1.4
Oleic C18:1	23.2
Linoleic C18:2	2.9
Linolenic C18:3	0.4
Total	96.9[a]

[a] The total weight % shown is not 100% because there are many other fatty acids (about 400) present in milk fat at very low concentrations.

12.2.1 The Copper Soap Solvent Extraction Method

A very sensitive assay is required since low concentrations of free fatty acids in milk have a marked impact on milk flavor. Moreover, milk is a very complex mixture so the assay must be quite specific for free fatty acids or other components may interfere. The assay we will use in this laboratory exercise is capable of detecting free fatty acids in milk at concentrations as low as $10\,mg\,l^{-1}$.

The Shipe assay doesn't measure individual fatty acids but rather provides an index of the extent of hydrolysis that has occurred by measuring free fatty acids (fatty acids not esterified to glycerol). Since it is difficult to measure free fatty acids directly, this assay takes advantage of the capacity of free fatty acids to form insoluble soaps with copper ions. These soaps along with their associated copper are extracted into a nonpolar solvent and the concentration of copper is determined. By using a standard curve, we can estimate the amount of free fatty acids present in the milk samples. The chemistry involved in the assay is summarized below.

In the first step, copper soaps of free fatty acids are formed by mixing Cu^{2+} with the milk (Figure 12.2). Copper is added as a complex with triethanolamine. The triethanolamine prevents the copper from precipitating in the milk.

The next step is to extract the copper soap with an organic solvent. Copper soaps are insoluble in water but soluble in nonpolar organic solvents. Shipe and his group determined through careful experimentation that a mixture of chloroform:heptane:methanol (CHM) in a ratio of 49:49:2 works best.

The final step is to measure the concentration of the copper (Cu^{2+}) in the CHM extract. Notice in Figure 12.2 that the relationship between the free fatty acids and the copper in the copper soap is stoichiometric (1 copper ion combines with 2 fatty acid molecules). Thus, if we know the concentration of copper in the extract, we also know the relative concentration of free fatty acids present. In order to make the determination of copper more sensitive, a color reagent is added, which combines with the copper to produce a yellow color that absorbs at 440 nm. The color reagent we will use is diethyldithiocarbamate (Figure 12.3).

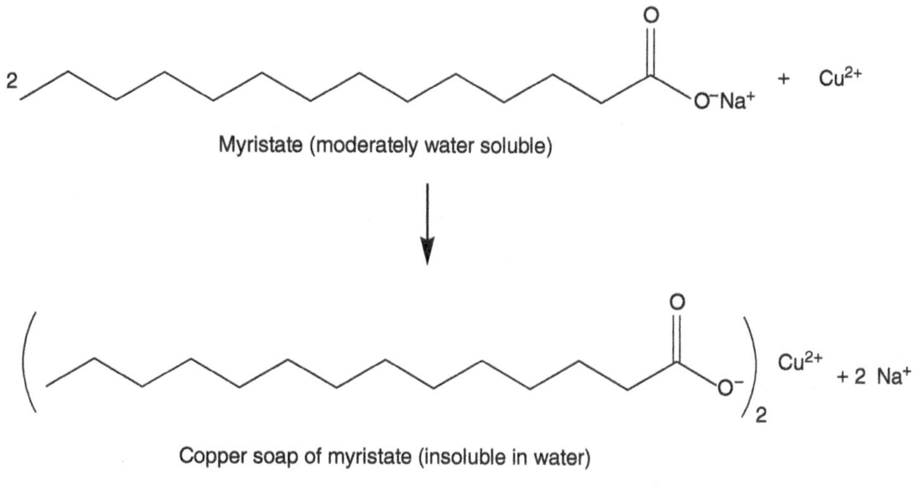

Myristate (moderately water soluble)

Copper soap of myristate (insoluble in water)

Figure 12.2 The formation of a copper soap in an aqueous system when Cu^{2+} is added to the solution. Note that there is a stoichiometric relationship between the Cu^{2+} and the fatty acid in the sample.

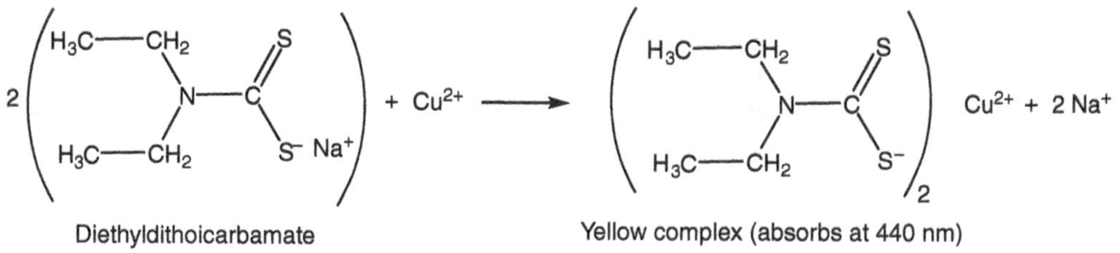

Diethyldithioicarbamate

Yellow complex (absorbs at 440 nm)

Figure 12.3 Cu^{2+} combines with diethyldithiocarbamate to form a complex that absorbs at 440 nm.

A standard curve is used to convert absorbances to free fatty acid concentrations.

12.3 Apparatus and Instruments

1) Blender
2) Flatbed reciprocating shaker
3) Visible spectrophotometer
4) Centrifuge
5) Vortex Mixer
6) Pipettes: 0.1, 0.5, 2, and 5 ml
7) Glass cuvettes (do not use disposable cuvettes, they dissolve in CHM).
8) Bottle top dispenser for dispensing CHM
9) Solvent-resistant screw-top centrifuge tubes (must be compatible with CHM).

12.4 Reagents and Materials

1) 0.7 N HCl
2) Copper reagent: 5 ml triethanolamine + 10 ml 1.0 M $Cu(NO_3)_2.3H_2O$ + saturated NaCl to 100 ml. The final pH should be adjusted to 8.3 with 1 N NaOH. Store in brown bottle.
3) Chloroform:heptane:methanol (CHM) 49:49:2; vol:vol:vol (be careful to mix thoroughly after combining the chloroform, heptane, and methanol).
4) Color reagent: 0.5% sodium diethyldithiocarbamate in n-butanol. (Sonicate to dissolve. Protect from light.)
5) Free fatty acid (FFA) standards: 0.4, 0.8, 1.2, and 1.6 mEq FFAl^{-1} CHM. (Note: FFA standards are a mixture of palmitic and myristic acids in a 3:2 molar ratio. Molecular weights are palmitic, 256.4; myristic, 228.4; recall that a milliequivalent of a fatty acid is equal to a millimole of the fatty acid since fatty acids are monoprotic, i.e. there is only one carboxyl group per molecule.)
6) Full-fat raw milk
7) Full-fat pasteurized-homogenized milk

12.5 Treatments and Controls

Treatment 1. Pasteurized–homogenized milk contaminated with raw milk. Pasteurized–homogenized milk spiked with raw milk (96 ml pasteurized–homogenized milk + 4 ml raw milk) and stored at 5 °C for 30 hours.

Treatment 2. Raw milk exposed to agitation. Raw milk (100 ml) blended in a food blender for 30 seconds and stored at 5 °C for 30 hours.

Treatment 3. Pasteurized–homogenized whole milk exposed to agitation. Pasteurized–homogenized whole milk (100 ml) blended in a food blender for 30 seconds and stored at 5 °C for 30 hours.

Control 1. Pasteurized–homogenized milk (100 ml) stored at 5 °C for 30 hours.

Control 2. Raw milk (100 ml) stored at 5 °C for 30 hours.

12.6 Procedures

12.6.1 Standard Curve

Prepare using standards containing palmitic and myristic acids mixed in a 3:2 molar ratio. *Run duplicate samples of each standard.*

1) Obtain 2 ml of each of the four standards (0.4, 0.8, 1.2, and 1.6 mEq FFA l^{-1} CHM).
2) Pipette 0.5 ml of distilled water into a 16 × 125 mm screw top centrifuge tube.
3) Pipette 0.1 ml of 0.7 N HCl into the tube. Mix on vortex mixer.
4) Pipette 2 ml of copper reagent into the tube and mix briefly using a vortex mixer.
5) Pipette 0.5 ml of standard into the tube.
6) Add 5.5 ml of CHM solvent and screw on the cap (Teflon lined); tighten securely.
7) Place tubes horizontally in shaker and shake on slow setting for 30 minutes. Be sure to place the tubes parallel to the motion of the shaker to ensure maximum mixing of the two phases.
8) Remove tubes from shaker and centrifuge for 10 minutes at 6000 rcf.
9) Transfer about 3.5 ml of the solvent layer (the top layer) to a clean, *dry* test tube (or directly into a cuvette). Be extremely careful to avoid getting any of the bottom layer into the transfer pipette. Even a small amount of the bottom layer will contain significant amounts of copper, and this will result in high readings.
10) Pipette 0.1 ml of color reagent into 3.5 ml from Step 9.
11) Mix and read on a spectrophotometer at 440 nm against a reagent blank (CHM solvent + color reagent). Use dry glass cuvettes (do not rinse with water).

12.6.2 Free Fatty Acids in Milk

1) Obtain 2 ml of each treatment and control. Do all determinations in duplicate.
2) Pipette 0.5 ml of milk sample into a 16 × 125 mm screw top centrifuge tube.
3) Pipette 0.1 ml of 0.7 N HCl into the test tube. Mix on vortex mixer.
4) Pipette 2 ml of copper reagent into same test tube and mix briefly using a Vortex mixer.
5) Add 6 ml of CHM solvent and screw on the cap (Teflon lined); tighten securely.
6) Place tubes horizontally in the shaker and shake on slow setting for 30 minutes.
7) Remove tubes from shaker and centrifuge for 10 minutes. After centrifuging, gently transfer the tubes to a test tube rack. Be careful to avoid mixing the two layers in the tubes when making this transfer.
8) Transfer about 3.5 ml of the solvent layer (the top layer) to a clean, *dry* test tube (or directly into a cuvette). Be extremely careful to avoid getting any of the bottom layer into the transfer pipette. Even a small amount of the bottom layer will contain significant amounts of copper, and this will result in high readings.
9) Pipette 0.1 ml of color reagent into 3.5 ml from Step 8.
10) Mix and read on a spectrophotometer at 440 nm against a reagent blank (CHM solvent + color reagent). Use dry glass cuvettes (do not rinse with water).

12.6.3 Calculations

Prepare a standard curve by plotting absorbance versus concentration of FFA in the standards. Express FFA concentration as mEq l^{-1}.

12.7 Problem Set

1 Given the following data, construct a standard curve and write the equation for the linear relationship between concentration and absorbance. Include the R^2 value for the relationship.

Free fatty acids (mEq l^{-1})	Absorbance (440 nm)
0	0
0.4	0.21
0.8	0.35
1.2	0.65
1.6	0.88

2 A milk sample was carried through the copper soap procedure as described above. The absorbance reading was 0.55. Using the data for the standard curve above, what was the concentration of free fatty acids in the milk, expressed as mEq l^{-1}?

12.8 Study Questions

1 What are the implications of your data for possible problems in the fluid milk industry?

2 Suggest approaches that would be effective in reducing the dairy industry's problem with rancid milk.

3 Suggest mechanisms and/or explanations for the effects of the various treatments on rancidity development.

4 Why is pasteurized milk less susceptible to rancidity development?

12.9 References

1 Santos, M.V., Ma, Y., Caplan, Z., and Barbano, D.M. (2003). Sensory threshold of off-flavors caused by proteolysis and lipolysis in milk. *Journal of Dairy Science* 86 (5): 1601–1607.
2 Yeung, C.K. (2020). Fatty acid composition of milkfat from dairy cows. Unpublished data.
3 Alvarez, V. (2009). Fluid milk and cream products. In: *The Sensory Evaluation of Dairy Products*, 2e (eds. S. Clark, M. Costello, M. Drake and F.W. Bodyfelt), 73–133. New York, NY: Springer.
4 Shipe, W.F., Senyk, G.F., and Fountain, K.B. (1980). Modified copper soap solvent extraction method for measuring free fatty acids in milk. *Journal of Dairy Science* 63 (2): 193–198.

12.10 Suggested Reading

Anderson, M. (1983). Milk lipase and off-flavour development. *International Journal of Dairy Technology* 36 (1): 3–7.

Elías-Argote, X., Laubscher, A., and Jiménez-Flores, R. (2013). Dairy ingredients containing milk fat globule membrane: description, composition, and industrial potential. In: *Advances in Dairy Ingredients* (eds. G.W. Smithers and M.A. Augustin), 71–98. Ames, Iowa: Wiley : Institute of Food Technologists. (IFT Press series).

Horn, D.S. (2017). Characteristics of milk. In: *Fennema's Food Chemistry*, 5e (eds. S. Damodaran and K.L. Parkin), 907–953. Boca Raton: CRC Press, Taylor & Francis Group.

Answers to Problem Set

1 Equation for standard curve: $y = 0.55x - 0.022$; $R^2 = 0.99$.

2 Concentration of free fatty acids in the milk sample $= 1.04\,mEq\,l^{-1}$.

13

Caffeine in Beverages

13.1   Learning Outcomes

After completing this exercise, students will be able to:

1) Measure the concentration of caffeine in selected beverages using a high-performance liquid chromatograph.
2) Explain the basic principles underlying liquid chromatographic methods.

13.2   Introduction

A large majority of American adults ingest caffeine on a daily basis. Most of us get our caffeine from coffee, tea, soda, and/or energy drinks. Caffeine is a central nervous system stimulant. It belongs to a class of compounds called methylxanthines, which are present naturally in over 60 species of plants, including coffee beans, tea leaves, cola nuts, and cocoa beans. In addition to being present in varying levels in the products made from these raw materials, caffeine is added as a flavoring agent in other foods and beverages, especially soft drinks and energy drinks. The structure of caffeine is shown in Figure 13.1.

Caffeine produces a range of physiological effects on the body. It may help keep you awake and enhance concentration, which is why many people, especially college students, feel they need their "caffeine fix" every morning. In addition to being a stimulant, it is a diuretic and may increase stomach acid secretion. It may increase blood pressure but the evidence on this is conflicting.

Numerous studies have demonstrated that caffeine enhances physical performance in a wide variety of athletic events [1]. Moreover, consumption of caffeine-containing beverages, especially coffee, is associated with reduced risk for some cancers, cardiovascular disease, type 2 diabetes, all-cause mortality, and possibly dementia [2]. Intervention studies on caffeine and physical performance often use purified caffeine, so the link between caffeine and physical performance is pretty clear. The health benefits of caffeine-containing beverages such as coffee and tea, however, may be due to polyphenols and other phytochemicals, not caffeine.

Too much caffeine may cause insomnia, restlessness, anxiety, diarrhea, headache, and heart palpitations. Caffeine affects people differently, so the amount associated with adverse effects will vary depending on a variety of factors including body weight, age, pregnancy, breast-feeding,

Food Chemistry: A Laboratory Manual, Second Edition. Dennis D. Miller and C. K. Yeung.
© 2022 John Wiley & Sons, Inc. Published 2022 by John Wiley & Sons, Inc.
Companion website: www.wiley.com/go/Miller/foodchemistry2

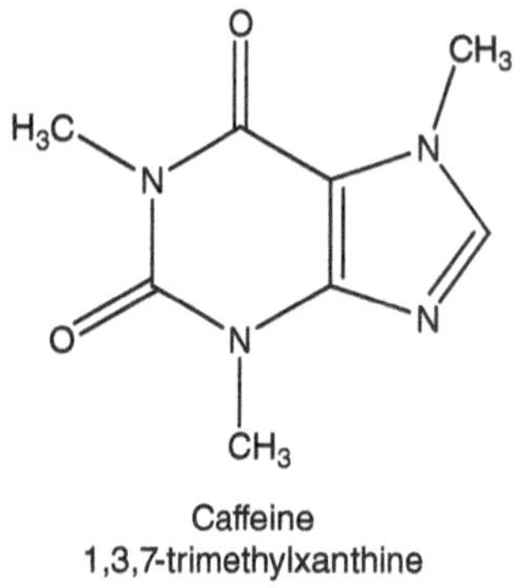

Figure 13.1 Caffeine is an alkaloid belonging to the class of methylxanthines. It has a slightly bitter taste.

Caffeine
1,3,7-trimethylxanthine

Table 13.1 Caffeine content in a typical serving of some common beverages [5, 6]

Beverage	Serving size (oz)	Caffeine content (mg)
Cola	12	30–40
Cola, decaffeinated	12	0
Citrus soft drinks (most brands)	12	0
Mountain Dew	20	91
Energy drinks	16	90–300
Coffee	8	80–100
Expresso	1	64
Tea, black	8	47
Tea, green	8	28

medications, and sensitivity to caffeine. Health experts generally recommend limiting caffeine intake to less than 400 mg a day, which is approximately the amount in four cups of brewed coffee. The American Academy of Pediatrics discourages consumption of caffeine-containing beverages for children under 12 and recommends that energy drinks containing caffeine and other stimulants "have no place in the diet of children and adolescents" [3]. The concentration of caffeine in energy drinks and energy shots varies widely, so it is advisable to check the label before consuming these products.

Because of perceived and real effects of caffeine on health and athletic performance, there has been a proliferation of energy drinks that contain high concentrations of caffeine. The consumption of these caffeinated energy drinks has been trending upward in adolescents and young adults [4]. Therefore, the concentration of caffeine in common beverages is an important variable for researchers and consumers alike. Two of these beverages, coffee and tea, are complex mixtures of many compounds that dissolve during the brewing or steeping process. Consequently, quantitation of a single compound in the mixture requires a separation step for accurate results. High-performance liquid chromatography (HPLC) has become the method of choice for the analysis of caffeine in beverages. See Appendix V for a discussion of HPLC theory.

The caffeine contents of several common beverages are listed in Table 13.1.

13.3   Apparatus and Instruments

1) HPLC unit equipped with a variable wavelength UV detector set at 254 nm
2) Reverse-phase C_{18} column
3) Guard column
4) Luer-lok syringe
5) Luer-lok filters, 0.45 μm (x10)
6) Vacuum aspirator
7) Sidearm flasks with rubber stoppers, 500 ml (x3)
8) Beakers, 250 ml (x2)
9) Sample vials, 8 ml (2/sample)
10) Sample vials, 4 ml (1/sample)
11) Micropipettes and tips
12) Sample injection syringe, 100 μl
13) Glass stirring rods (x2)
14) Hot plate
15) Flask, 500 ml
16) Gloves or tongs
17) Analytical balance, spatula, and weighing paper
18) Solvent filtering apparatus

13.4   Reagents and Materials

1) Caffeine standard solutions: 25–250 ppm, to be prepared by the teaching staff. KEEP REFRIGERATED UNTIL READY TO USE!!
2) HPLC mobile phase:
    i) HPLC grade methanol, 350 ml
    ii) HPLC grade water, 650 ml
    iii) Acetic acid, 4 ml
  Combine, then filter through a 0.45 um Nylon-66 filter, and de-gas for five minutes.
3) Syringe rinse (HPLC grade methanol), in 8 ml sample vial.
4) Samples for analysis:
    i) Diet cola, caffeinated
    ii) Diet cola, decaffeinated
    iii) Energy drink, caffeinated
    iv) Coffee, brewed or instant
    v) Coffee, brewed or instant, decaffeinated
    vi) Black tea, brewed
    vii) Green tea, brewed

13.5   Operation of the HPLC

(**Note**: the operation parameters will vary depending on the make and model of the HPLC unit. Carefully follow the instructions for operation in the manual supplied by the manufacturer.)

1) Start-Up
   a) Turn on the HPLC unit.
   b) Allow to warm up for 30 minutes.
2) Sample Preparation
   a) Place a 0.45 μm filter on the Luer-lok syringe. Remove the plunger.
   b) Pour approximately 3 ml of the sample to be tested from a sample vial into the syringe, and replace the plunger.
   c) Carefully filter the sample into a small sample vial.
   d) Remove the filter and rinse the syringe with HPLC grade water. (Install a new filter for each filtration.)
3) Injection and Run
   a) Zero the integrator.
   b) Rinse the sample injection syringe five times with the sample, emptying the syringe into the waste bucket after each rinse.
   c) Rinse the syringe five more times with the sample by slowly pumping the sample in and out of the syringe.
   d) Carefully fill the syringe to its set volume (70 μl), making sure no air bubbles are trapped in the barrel.
   e) Adjust the syringe volume to 60 μl, blotting the tip with a Kim-wipe.
   f) Rotate the injector to the "LOAD" position.
   g) Insert the syringe into the injector.
   h) Press the plunger quickly and carefully, then turn the injector to "INJECT" while pressing the "START" button on the integrator.
   i) Remove the syringe from the injector.
   j) Rinse the syringe five times with HPLC grade methanol.
   k) After the last peak has eluted from the column, press the "STOP" button on the integrator.
   l) Rinse the injector between samples by injecting, in the load position, three syringe loads of HPLC grade methanol.
4) Interpretation of Results
   a) The integrator will provide a chromatogram and a listing of the retention times and areas of the peaks.
   b) Find the caffeine peaks by comparing the retention times of the samples with those of the standards.
   c) Using the standard curve, compute the concentration of caffeine in the sample. Express the results as mg caffeine/6 oz. (180 ml) serving.
5) Shutdown
   a) Reduce the flow rate to 0.06 ml min^{-1}.
   b) Flush the injector three times with 100% HPLC grade methanol while it is in the "LOAD" position.

13.6 Procedures

13.6.1 Standard Curve

1) A standard curve will be prepared using the caffeine standards. Plot average peak area vs. caffeine concentration in mg ml^{-1}. Remember that 1 ppm = 1 mg l^{-1}. The actual standards will be run by the individual student groups.

13.6.2 Caffeine in Soda and Energy Drinks

1) Place approximately 200 ml of the soda to be tested in a 500 ml sidearm flask and stopper it.
2) Attach the flask to the vacuum pump and aspirate for 5–10 minutes, or until no more bubbles are present, while swirling the flask. Remove the hose *before* turning off the pump.
3) With a 10 ml pipette, transfer 5 ml de-gassed soda into an 8 ml sample vial.
4) Prepare and inject samples as described in Section 13.5. Each sample should be run two times to minimize error.

13.6.3 Caffeine in Coffee

1) Boil approximately 250 ml dH$_2$O in the 500 ml flask.
2) Place 2.15 g (approximately one rounded teaspoon) instant coffee in a clean 250 ml beaker.
3) Add 180 ml (6 ounces) boiling water to the beaker. Mix with a glass stirring rod until dissolved.
4) With a 10 ml pipette, transfer 5 ml coffee into an 8 ml sample vial. Coffee containing caffeine should be diluted 1:3 with dH$_2$O (1 part coffee to 2 parts dH$_2$O), in another 8 ml sample vial. Decaffeinated coffee should not be diluted.
5) Prepare and inject samples as described in Section 13.5. Each sample should be run two times to minimize error.

13.6.4 Caffeine in Tea

1) Prepare as with coffee.

13.7 Data Analysis

1) Plot a standard curve using the chromatograms prepared in class. Plot average peak area vs. caffeine concentration in mg ml^{-1}.
2) Calculate mg caffeine/6 oz. serving for each beverage using standard curve results. Remember that the coffee has been diluted.
3) Assuming that the decaffeinated coffee was originally identical to the coffee with caffeine, calculate the percent decaffeination.

13.8 References

1 Pickering, C. and Grgic, J. (2019). Caffeine and exercise: what next? *Sports Medicine* 49 (7): 1007–1030.
2 Grosso, G., Godos, J., Galvano, F., and Giovannucci, E.L. (2017). Coffee, caffeine, and health outcomes: an umbrella review. *Annual Review of Nutrition* 37 (1): 131–156.
3 Heckman, M.A., Weil, J., and Mejia, E.G.D. (2010). Caffeine (1, 3, 7-trimethylxanthine) in foods: a comprehensive review on consumption, functionality, safety, and regulatory matters. *Journal of Food Science* 75 (3): R77–R87.
4 Vercammen, K.A., Koma, J.W., and Bleich, S.N. (2019). Trends in energy drink consumption among U.S. Adolescents and adults, 2003–2016. *American Journal of Preventive Medicine* 56 (6): 827–833.
5 FDA (2019). Spilling the beans: how much caffeine is too much? [cited 30 April 2020]. https://www.fda.gov/consumers/consumer-updates/spilling-beans-how-much-caffeine-too-much
6 Mayo Clinic. How much caffeine is in your cup? [Internet]. [cited 30 April 2020]. https://www.mayoclinic.org/healthy-lifestyle/nutrition-and-healthy-eating/in-depth/caffeine/art-20049372

13.9   Suggested Reading

DiNunzio, J.E. (1985). Determination of caffeine in beverages by high performance liquid chromatography. *Journal of Chemical Education* 62 (5): 446.

Nielsen, S.S. and Talcott, S.T. (2017). High-performance liquid chromatography. In: *Food Analysis Laboratory Manual*, 3e (ed. S.S. Nielsen), 77–85. Cham: Springer.

Strohl, A.N. (1985). A study of colas: an HPLC experiment. *Journal of Chemical Education* 62 (5): 447.

14

Color Additives

14.1 Learning Outcomes

After completing this exercise, students will be able to:

1) Extract, concentrate, and identify color additives in a food product.
2) Describe the chemical principles that underlie the processes in Outcome 1.
3) Use solid-phase extraction (SPE) to separate red dye from other colorants in colored beverages and quantify the amount present.

14.2 Introduction

Color is an important sensory aspect of most foods. We often use color as an index of freshness, wholesomeness, and overall quality. Unfortunately, the color of a food may change during processing, storage, or preparation in ways that are often perceived as undesirable. Some foods (e.g. cola drinks and gelatins) are colorless unless a colorant is added, while other foods may be made more appealing by enhancing or changing the natural color. Thus controlling, changing, and/or stabilizing the color of foods is a major objective for food scientists and technologists.

In the eighteenth and nineteenth centuries, food manufacturers used many different chemicals to color foods. Pickles were colored green with copper sulfate. Candy was colored with salts of copper and lead. Today, many artificial food colors are chemically synthesized from organic compounds. These artificial colorants were first synthesized from by-products of coal production, hence the term "coal tar dyes." Modern synthetic food colorants are made from purified petrochemicals, so the term "coal tar dye" is no longer accurate, but the phrase persists in the popular literature.

Food colorants have been controversial almost since they were first introduced because they have the reputation of being potentially toxic, because they may be used to deceive the consumer, and because their primary function is to enhance appearance rather than nutritive value, shelf life, or safety. With the exceptions of riboflavin, beta-carotene, and apo-carotenal, color additives have no nutritional value.

Today, both natural and synthetic colorants are added to foods. The natural pigments include annatto extract, dehydrated beets, β-carotene, paprika, and many others [1]. The number of synthetic food colorants approved for use has declined over the years as government safety regulations and toxicity testing have advanced. Also, many food manufacturers are moving away from synthetic colorants in favor of "natural" colorants to achieve "clean labels" on their products.

Food Chemistry: A Laboratory Manual, Second Edition. Dennis D. Miller and C. K. Yeung.
© 2022 John Wiley & Sons, Inc. Published 2022 by John Wiley & Sons, Inc.
Companion website: www.wiley.com/go/Miller/foodchemistry2

Table 14.1 FDA "certified" synthetic color food additives [4].

FD&C number	Common name	Chemical class	Characteristics	Uses and restrictions
FD&C Blue No. 1	Brilliant blue	TPM	Stable to heat, unstable to light	Foods generally
FD&C Blue No. 2	Indigotine	Indigoid	Stable to light, unstable in water	Foods generally
FD&C Green No. 3	Fast green	TPM	Teal green	Foods generally
FD&C Red No. 3	Erythrosine	Xanthene	Stable to heat, unstable to light	Foods generally
FD&C Red No. 40	Allura red	Azo	Unstable to redox agents	Foods generally
FD&C Yellow No. 5	Tartrazine	Azo	Good heat and light stability	Foods generally
FD&C Yellow No. 6	Sunset yellow	Azo	Fair heat and light stability	Foods generally
	Orange B	Azo		Use restricted to sausage casings but has not been used in recent years
	Citrus Red No. 2	Azo		Use restricted to skins of oranges grown in Florida

The FDA regulates food color additives under the authority of the 1960 Color Additives Amendment to the Federal Food, Drug, and Cosmetic Act of 1938 [2]. Food colorants are classified by the FDA as either "certified" or "exempt from batch certification."

The certified colors are synthetic organic compounds. Manufacturers must submit a sample from each production batch to the FDA for certification. Currently, there are nine certified color additives that may be added to foods (Table 14.1). Most of the certified colorants used in the United States have been assigned an FD&C number as required by the Food, Drug and Cosmetic Act. These numbers were established to avoid confusion between colorants manufactured for food use and colorants manufactured for other purposes. The same colorant may be used in both cases, but the food colorant must be free of toxic contaminants. Orange B, which is approved for sausage casings, is FDA certified but has not been used in several years [3]. Citrus Red No. 2 is also certified, but its use is restricted to the skins of oranges.

Most of the 29 "exempt from batch certification" colors for foods are extracted from natural sources, usually plants. A few synthetic dyes that are "nature-identical," for example synthetic β-carotene, are also exempt (see Chapter 15).

Most of the FD&C colorants are sodium salts of sulfonic acids. Structures of these colorants are shown in Figure 14.1.

Food colorants are added to foods at low concentrations. Consequently, it is often necessary to extract *and* concentrate the colorants in order to obtain sufficient amounts for analysis. In many cases, a procedure for separating and concentrating substances in foods can be developed from knowledge of the properties imparted by various functional groups in the molecules of interest. Note that the colorants shown in Figure 14.1 contain sulfonate groups, i.e. they are salts of sulfonic

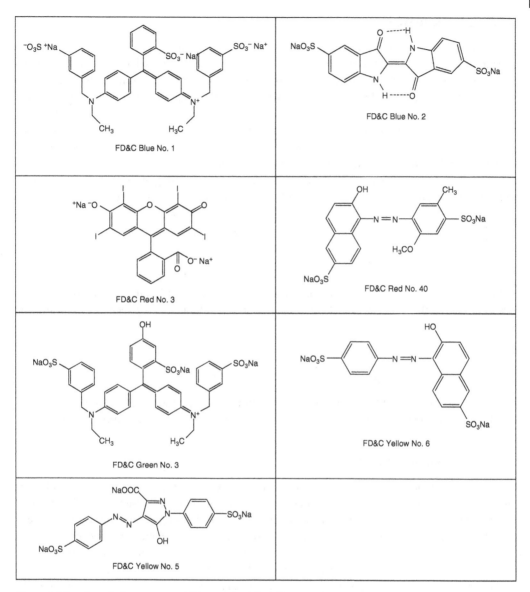

Figure 14.1 Chemical structures of FDA certified FD&C food colorants [5].

acids. Sulfonic acids are relatively strong acids. Thus, we expect FD&C colorants will be ionized and carry negative charges at pH values found in most foods. This ionic character makes them water soluble and provides a means for separating them from other components in the food.

In this experiment, extraction will be accomplished two ways.

1) *Binding the colorants to wool* and later releasing them into an aqueous solution. Principles underlying this extraction are summarized below.
2) *Solidphase extraction (SPE)*. SPE is similar to binding to wool (both involve binding colorants to a solid phase) but uses a more sophisticated solid phase.

14.2.1 Binding to Wool

We will take advantage of electrostatic interactions between the colorant molecules and a protein to separate and concentrate the colorants. Wool (a strand of wool yarn) will be used as the protein. Wool protein is a suitable binding agent for this purpose because it is insoluble and because its charge can be manipulated by changing the pH. At low pH, carboxyl and amino groups on the wool protein will be protonated, giving the wool protein a net positive charge. The colorant molecules, on the other hand, remain negatively charged at low pH because they are salts of a strong acid. Acetic acid, a weak acid, will be used to acidify the food so that when the wool is added, it will be protonated. Electrostatic bonding between positively charged protein molecules and negatively charged colorant molecules probably accounts for most of the bonding of the colorant to the wool strand although some hydrogen bonding and hydrophobic interactions may be involved as well:

| Dye molecule | $-SO_3^-$ ---------- $^+H_3N-$ | Wool protein |

(Electrostatic bonding)

When the colorant molecules are bound to wool under these conditions, rinsing in cold water does not remove the color. This indicates a fairly strong interaction between the colorant and the wool.

14.2.2 Removal from Wool

Gentle boiling of the wool fiber in dilute alkali deprotonates the amino groups on the wool protein. This disrupts electrostatic bonding forces, and the color is released from the strand into solution.

14.2.3 Solid-Phase Extraction (SPE)

SPE has largely replaced liquid/liquid extraction as a method for extracting and concentrating solutes in liquid solutions. In liquid/liquid extractions, a liquid that is immiscible with the liquid containing the solute (the sample) is shaken with sample. For example, one might use hexane, a nonpolar hydrocarbon, to extract a colorant with some nonpolar character from a fruit juice sample. The juice and the hexane would be placed in a separatory funnel. After shaking vigorously, the two immiscible layers would be allowed to separate by gravity and the lower aqueous layer would be drained through a valve in the bottom of the separatory funnel leaving the hexane with the dissolved colorant behind. This technique requires a large volume of solvent and the separation is rarely complete.

In SPE, a solid packing material (usually some form of silica) to which polar, nonpolar, or ionic molecules are covalently bound is placed in a column (or a syringe), and the liquid sample is applied to the top of the packing. There are three main types of packing: normal phase, reversed phase, and ion exchange. In normal phase, the packing material (stationary phase) is polar, and the liquid phase is nonpolar. In reversed phase, the stationary phase is nonpolar and the liquid phase is polar. In ion exchange, the stationary phase contains charged groups and the liquid phase is usually polar. The wool extraction described above is an example of an ion exchange SPE. In this experiment, we will be using reversed-phase SPE to extract FD&C Red No. 40 from cranberry juice. The packing material will be silica to which an 18-carbon hydrocarbon is covalently linked, hence the name Bond Elut C-18 (Figure 14.2).

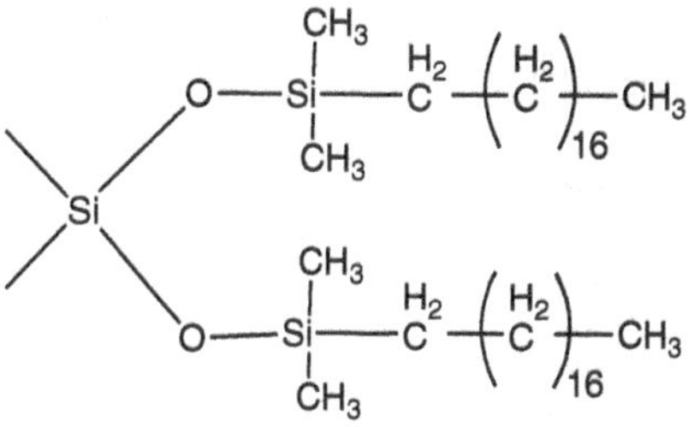

Figure 14.2 Structure of a C-18 bonded silica packing material. Note that the silicon atoms on the surface of the silica are covalently linked to 18-carbon hydrocarbon chains. This material is highly nonpolar and will attract nonpolar solutes.

14.2.4 Separation and Identification

Separation and identification will be accomplished by thin layer chromatography (TLC). TLC is a form of adsorption chromatography. Separation is accomplished by the differential adsorption of components in the sample on a stationary phase. Samples and standards are applied at the bottom of a plate coated with silica gel. The plate is then placed vertically in a chamber containing a small amount of solvent in the bottom, taking care that the solvent does not reach up to the sample. As the solvent system moves up the plate by capillary action and moves through the sample, components of the sample, which are soluble in the solvent, are carried upward with the solvent. Since some components are adsorbed more strongly than others, they begin to separate. More polar compounds adsorb more strongly and remain nearer the origin. Less polar compounds adsorb only weakly and thus spend more time in the moving solvent, thus migrating faster. When the solvent front reaches the top of the plate, the plate is removed and examined. By comparing the colors and migration distances of the samples and standards, it is often possible to definitively identify components present in the samples. The positions of compounds on a TLC plate are often described by the R_f value:

$$R_f = \frac{d_{compound}}{d_{solvent}}$$

where: $d_{compound}$ = distance traveled by the compound
$d_{solvent}$ = distance traveled by the solvent front

If identical solvents and stationary phases are used, the R_f value for a particular compound in that system should remain constant. Table 14.2 lists R_f values for colorants run on different silica gel plates. Note that R_f values for the same compound differ between the plates but that the order (lowest to highest R_f value) are similar for the plates. Do you have an explanation?

Note: The FDA has published methods for *quantitative* analysis of certified color additives in food products. In these methods, the color additives are extracted using methanol/aqueous ammonium hydroxide as the solvent. The extracts are analyzed using reversed-phase liquid chromatography [7].

14.3 Apparatus and Instruments

1) Beakers, 50, 100, and 200 ml
2) Graduate cylinders, 10 and 50 ml
3) Boiling beads
4) Pipettors with tips
5) Oven set at 95 °C
6) Hot plates
7) Pre-coated silica gel plastic plates (plates should be activated by heating for four hours at 50 °C prior to use). **Note:** always check manufacturer's instructions before using pre-coated plates.

Table 14.2 R_f values[a] × 100 for synthetic food colors chromatographed on TLC plates prepared by different manufacturers[b] [6].

Colorant	FD&C #	R_f × 100[c]	R_f × 100[d]
Allura red	Red No. 40	No value	24–38
Tartrazine	Yellow No. 5	36–39	17–20
Sunset yellow	Yellow No. 6	47–48	32–36
Fast green FCF	Green No. 3	29–32	15–16
Brilliant blue	Blue No. 1	39–41	20–25
Indigotine	Blue No. 2	37–40	26–27

[a] Measured to the leading edge of the spot.
[b] Isopropanol/concentrated ammonia (4:1) was the developing solvent.
[c] Baker-flex silica gel plates.
[d] Machinery-Nagel glass plates coated with silica gel.

8) Hair dryer
9) Developing tank
10) C-18 column (Agilent Bond Elut JR-C18, 500 mg cartridge or similar)
11) UV/Vis spectrophotometer
12) Cuvettes
13) Syringes, 10 ml
14) Vials (pre-weighed), 10 or 20 ml

14.4 Reagents and Materials

1) Assorted colored gelatins and canned sodas
2) FD&C Colors for standards: Red No. 3 and 40, Blue No. 1 and 2, Green No. 3, and Yellow No. 5 and 6.
3) White knitting wool yarn (purified in advance by boiling in 0.01 N NaOH and then boiling in water)
4) Acetic acid, 5 N
5) Ammonium hydroxide, 0.5 N
6) Isopropanol/concentrated ammonia (4:1, v/v)
7) Ethanol, 95%
8) FD&C Red No. 40 stock solutions (ranging 1–15 µg ml^{-1}) for SPE experiment
9) Commercial cranberry juice containing Red No. 40 or cranberry juice spiked with Red No. 40
10) 2-propanol (4 and 5% v/v in water)
11) Acidified methanol (MeOH) (add 1 ml concentrated HCl to 99 ml methanol)

14.5 Procedures

14.5.1 Qualitative Identification of Artificial Colors from Food Products

1) Obtain one gelatin sample and one soft drink sample.
2) Transfer a 50 ml aliquot of the soft drink to a 100 ml beaker and acidify with 1 ml of 5 N acetic acid.

3) Transfer 2.5 g of gelatin to a 100 ml beaker. Dissolve in 50 ml of water and acidify with 1 ml of 5 N acetic acid.
4) Drop a 20 cm strip of white knitting wool (purified in advance by the instructor) into each acidified sample. Add boiling beads. Boil the mixtures gently for three to five minutes. Cool to room temperature.
5) Wash the wool with cold water. Transfer it to a small beaker. Add boiling beads and about 10 ml of 0.5 N ammonia. Boil gently until the color is released into solution.
6) After the color is released, discard the wool and put the solution in a 95 °C oven until it reaches a state of near dryness. Alternatively, the water can be evaporated on a hot plate. (If you choose to use the hot plate, **USE CAUTION.** Hot solution may spatter out of the beaker.)

14.5.2 Separation and Identification of the Extracted Colors

1) Spot 10–20 µl of each FD&C colorant and your extracts on silica gel plates. Spots should be at least 2 cm from the bottom of the plate and no more than 0.5 cm in diameter. Dry the spots by gently heating with a hair dryer. Each group will receive one plate, and each student should spot their sample plus at least two FD&C food colorants on a plate. That way, everyone will gain experience in spotting thin layer plates. All standards must be run on each plate. The food colorant standards can be spotted directly (5 µl) without dilution. Use a maximum of nine samples per plate.
2) When all of the samples and standards have been spotted on the plate, transfer it to a developing tank containing the mobile phase [isopropanol/concentrated ammonia (4:1, v/v)]. Allow the plates to develop until the solvent front is 2–4 cm from the top of the plate.
3) Calculate R_f values for all spots and compare known FD&C standards with unknowns in food products for tentative identification.
4) Compare your R_f values with values in Table 14.2.

14.5.3 Quantitative Analysis of FD&C Red Dye # 40 in Cranberry Juice

Note: This section was developed and written by Professor Chang Yong Lee. It was adapted from a procedure described by Rossi et al. [8].

1) Standard curve: Measure the absorbance of Red No. 40 standards (0, 1 µg, 5 µg, 10 µg, 15 µg ml^{-1}) at 502 nm and prepare a standard curve (plot absorbance vs. concentration).
2) Separation and quantification of Red No. 40 using SPE:
   a) Condition the column with 5 ml of acidified methanol. To do this, attach a syringe containing 5 ml methanol to the column and gently force the methanol through the column by applying pressure to the syringe plunger.
   b) Equilibrate the column with 5 ml of deionized water.
   c) Load the column with 1.0 ml of beverage.
   d) Wash the column three times with the reagents shown below. Combine these three washes in a pre-weighed vial.
      i) Wash 1: 3 ml of 4% 2-propanol
      ii) Wash 2: 3 ml of 4% 2- propanol
      iii) Wash 3: 3 ml of 5% 2-propanol
      iv) Weigh the vial and calculate the weight of the combined washes. Calculate the volume of the combined washes (since the washes are > 95% water, we can assume the density of the solutions is 1 g ml^{-1}).

e) Elute the column with 5 ml of acidified methanol. Anthocyanins will be washed off the column and the column is ready for reuse.
f) Measure the absorbance of the washes in the weighed vial at 502 nm.
g) Calculate the concentration of Red No. 40 in the **original juice**.

14.6 Study Questions

1 Compare your R_f values (from your standards) with those in Table 14.2. Are they similar in size? In order? Explain the differences among the three sets of R_f values.
2 Rank the pK values for the following acids: acetic, sulfonic, $R\text{-}NH_3^+$ where $-NH_3^+$ is the ε-amino group on a lysine residue in a protein. Explain, on the basis of this ranking, why the extraction procedure with wool yarn used in this experiment worked.
3 Explain why the Red No. 40 was eluted from the column with 3-propanol, while the anthocyanins were retained. Why did the acidified methanol elute the anthocyanins?

14.7 References

1 Sigurdson, G.T., Tang, P., and Giusti, M.M. (2017). Natural colorants: food colorants from natural sources. *Annual Review of Food Science and Technology* 8: 261–280.
2 Schwartz, S.J., Cooperstone, J.L., Cichon, M.J. et al. (2017). Colorants. In: *Fennema's Food Chemistry*, 5e (eds. S. Damodaran and K.L. Parkin), 681–752. Boca Raton: CRC Press, Taylor & Francis Group.
3 Kobylewski, S. and Jacobson, M.F. (2012). Toxicology of food dyes. *International Journal of Occupational and Environmental Health* 18 (3): 220–246.
4 FDA (2020). Summary of color additives for use in the United States in foods, drugs, cosmetics, and medical devices. [cited 17 March 2020]. http://www.fda.gov/industry/color-additive-inventories/summary-color-additives-use-united-states-foods-drugs-cosmetics-and-medical-devices
5 Brady, J.W. (2013). *Introductory Food Chemistry*, 638. Ithaca: Comstock Publishing Associates.
6 Dixon, E.A. and Renyk, G. (1982). Isolation, separation and identification of synthetic food colors. *Journal of Chemical Education* 59 (1): 67–69.
7 Petigara Harp, B., Miranda-Bermudez, E., and Barrows, J.N. (2013). Determination of seven certified color additives in food products using liquid chromatography. *Journal of Agricultural and Food Chemistry* 61 (15): 3726–3736.
8 Rossi, H.F., Rizzo, J., Zimmerman, D.C., and Usher, K.M. (2012). Extraction and quantitation of FD&C red dye #40 from beverages containing cranberry juice: a college-level analytical chemistry experiment. *Journal of Chemical Education* 89 (12): 1551–1554.

14.8 Suggested Reading

Frick, D. and Huck, P. (1995). Food color terminology. *Cereal Foods World* 40 (4): 209–218.
Hallagan, J.B. (1991). The use of certified food color additives in the United States. *Cereal Foods World* 36 (11): 945–948.
McKone, H.T. and Nelson, G.J. (1976). Separation and identification of some FD&C dyes by TLC. An undergraduate laboratory experiment. *Journal of Chemical Education* 53 (11): 722.
Supelco. Guide to solid phase extraction [Internet]. [cited 18 March 2020]. https://www.sigmaaldrich.com/Graphics/Supelco/objects/4600/4538.pdf

15

Plant Pigments

15.1 Learning Outcomes

After completing this exercise, students will be able to:

1) Describe the major classes of plant pigments and their distributions in foods.
2) Extract water soluble and fat soluble pigments from plants.
3) Separate extracted pigments using thin layer chromatography (TLC).

15.2 Introduction

The appealing colors of fruits and vegetables are due to compounds in the plant tissues that absorb light of certain wavelengths. These pigments make up a large, structurally diverse group of compounds whose presence and relative concentrations vary with plant species, degree of ripeness, and growing conditions. They may be classified into two groups based on structure: compounds containing *conjugated double bonds* and compounds with *metal-coordinated porphyrin rings* (which also contain conjugated double bonds). Carotenoids and anthocyanins fall into the first group while chlorophylls belong to the second.

Recall that conjugated double bonds are double bonds separated by a single bond as shown in Figure 15.1.

Many of the naturally occurring plant pigments are available in purified, concentrated form for use as color additives. They fall into the group of colorants classified by the FDA as "color additives exempt from certification." Currently, 29 "color additives exempt from certification" are approved for use in human foods (Table 15.1). Some of them, while naturally occurring, are available in synthetic form. In most cases, the synthetic form is chemically identical to the naturally occurring form and therefore is called "nature identical." In general, naturally occurring colorants are more expensive and less stable than the synthetic FD&C colorants. Thus, many of the 29 colorants in this group are rarely used.

Anthocyanins are water soluble compounds ranging in color from deep purple to orange-red. They are found primarily in fruits but are also present in some vegetables, e.g. radishes, eggplant, red cabbage, and red potatoes. Anthocyanins are glycosides of anthocyanidins (Figure 15.2).

The color of an anthocyanins is very sensitive to pH. They tend to be red in acid, colorless around pH 4, and blue in the neutral pH range (Figure 15.3).

Food Chemistry: A Laboratory Manual, Second Edition. Dennis D. Miller and C. K. Yeung.
© 2022 John Wiley & Sons, Inc. Published 2022 by John Wiley & Sons, Inc.
Companion website: www.wiley.com/go/Miller/foodchemistry2

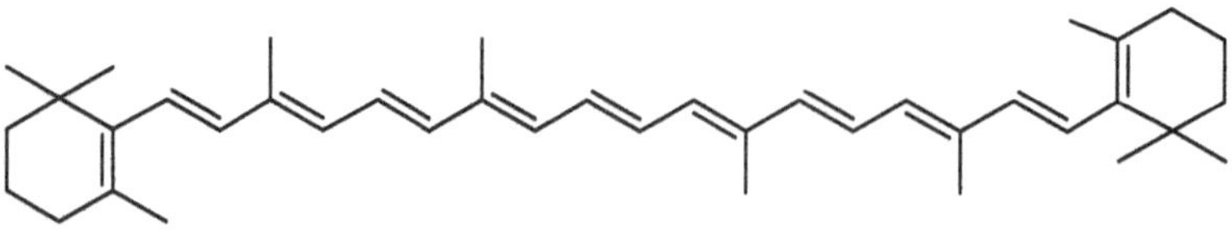

Figure 15.1 The structure of β-carotene, an example of a molecule containing conjugated double bonds. Note that single and double bonds alternate in the structure.

Table 15.1 FDA approved "color additives exempt from batch certification" for use in human foods [1, 2].

Color additive: color	Source	Uses
Annato extract: yellowish-orange	Seeds of *Bixa orellana* tree	Cheddar cheese and foods generally
Dehydrated beets: red	Beets	Foods generally
Calcium carbonate: white	Synthesis, or naturally occurring limestone	Soft and hard candies and mints
Canthaxanthin: reddish orange	Synthesis, or mushrooms and trout and salmon	Foods generally
Caramel: brown to dark brown	Heating corn syrup	Foods generally
β-apo-8'-carotenal: reddish orange	Citrus fruit skin, vegetable pulp	Foods generally
β-carotene: orange	Synthesis, or algae	Foods generally
Carrot oil: yellowish orange	Carrots	Foods generally
Cochineal/carmine: red	Female cochineal insect	Foods generally
Sodium copper chlorophyllin: green	Chlorophyll extracted from alfalfa	Citrus-based dry beverage mixes
Cottonseed flour: brown	Toasted cottonseed	Foods generally
Ferrous gluconate: color fixation in ripe black olives	Synthesis from iron and gluconic acid	Ripe olives
Ferrous lactate: color fixation in ripe black olives	Synthesis from iron and lactic acid	Ripe olives
Grape color extract: reddish purple	Concord grapes	Nonbeverage food
Grape skin extract: reddish purple	Grape skins	Beverages, alcoholic beverages
Iron oxide: yellow, red, brown and black	Synthesis	Sausage casings, hard and soft candy, mints and chewing gum
Fruit juice: various color	Variety of fruits	Foods generally
Vegetable juice: various color	Variety of vegetables	Foods generally
Mica-based pearlescent pigments: various pearlescent color effects	Synthesis	Cereals, frostings, candies
Paprika: red-orange	Sweet red peppers	Foods generally
Paprika oleoresin: red-orange	Sweet red peppers	Foods generally
Riboflavin: yellow	Synthesis	Foods generally
Saffron: yellow	*Crocus sativus* plant	Foods generally
Soy leghemoglobin: reddish brown	Controlled fermentation of genetically engineered yeast, *Pichia pastoris*	Ground beef analogue products
Spirulina extract: blue	Aqueous extraction of the dried biomass of *Arthrospira platensis*	Candy and chewing gum, frostings, etc.
Tomato lycopene extract; tomato lycopene concentrate: red	Tomato	Foods generally
Titanium dioxide: white	Ilmenite (a mineral oxide)	Foods generally
Turmeric: yellow	*Curcuma longa L.* plant	Foods generally
Turmeric oleoresin: yellow	*C. longa L.* plant	Foods generally

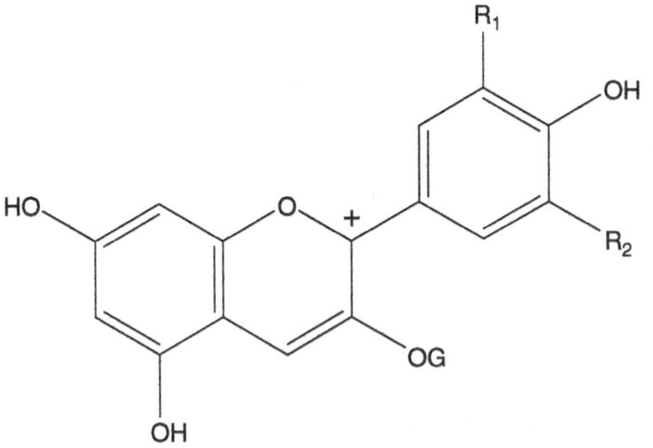

Figure 15.2 Generalized structure of an anthocyanin. G is a glucose residue. R_1 and R_2 groups may be H, OH, or OCH_3. R_1 and R_2 groups vary among anthocyanins from different sources. A single fruit or vegetable may contain more than one form.

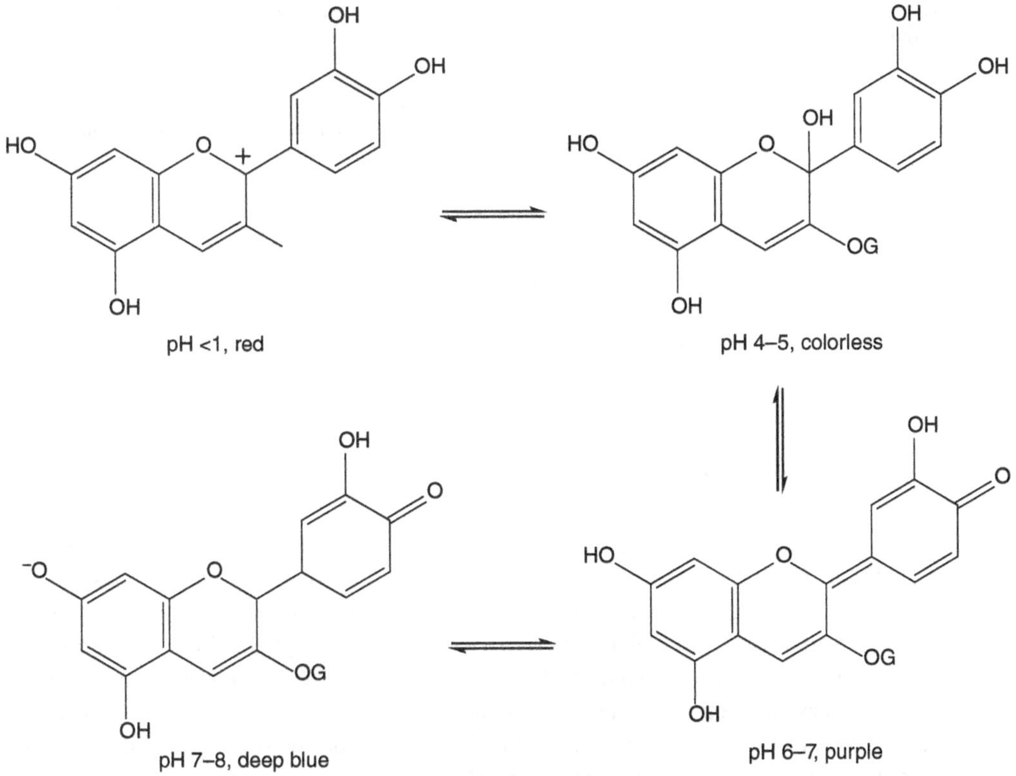

pH <1, red

pH 4–5, colorless

pH 7–8, deep blue

pH 6–7, purple

Figure 15.3 Color of cyanin, an anthocyanin, at various pHs. G is a glucose residue. Redrawn from [3].

Chlorophylls are green pigments containing a porphyrin ring complexed with magnesium. The two major chlorophylls, a and b, differ in that a methyl group in chlorophyll a is replaced by an aldehyde group in chlorophyll b (Figure 15.4). Chlorophyll a and b are degraded to pheophytin a and b, respectively, by the replacement of magnesium with two protons. This degradation alters the color of green-colored plants from bright green to dull olive brown. Acid conditions produced during thermal processing often cause this color change, especially in canning.

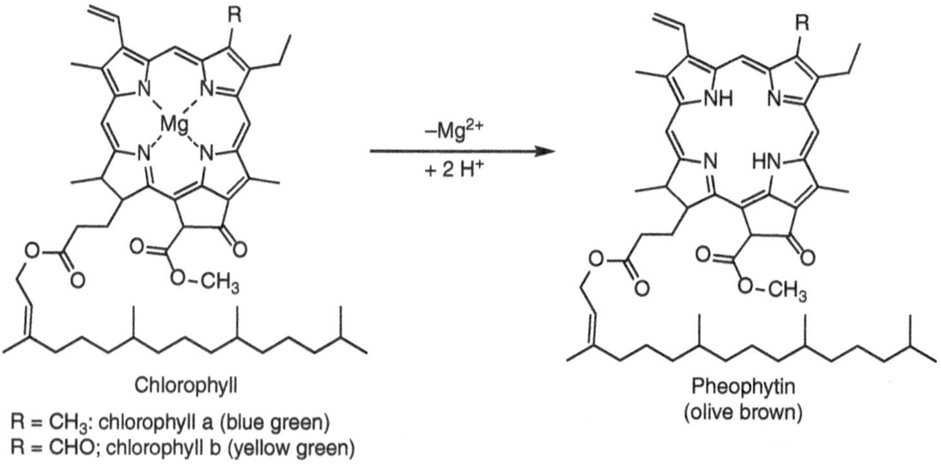

Chlorophyll

$-Mg^{2+}$
$+ 2 H^+$

Pheophytin
(olive brown)

R = CH₃: chlorophyll a (blue green)
R = CHO; chlorophyll b (yellow green)

R = CH_3: chlorophyll a (blue green)
R = CHO; chlorophyll b (yellow green)

Figure 15.4 Conversion of chlorophyll to pheophytin. Chlorophyll is called a magnesium porphyrin because the four nitrogen atoms in the porphyrin ring are coordinated with a magnesium ion. Chlorophylls a and b occur together in about a 3:1 ratio. Chlorophyll a is blue green, while chlorophyll b is yellow green. Redrawn from [4].

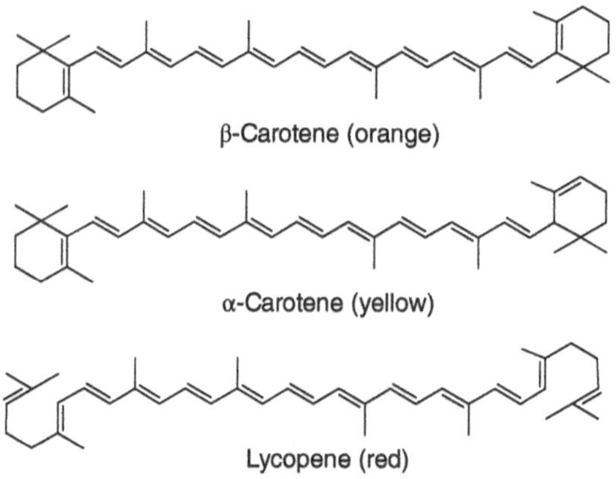

β-Carotene (orange)

α-Carotene (yellow)

Lycopene (red)

Figure 15.5 Structures of selected carotenes. Note the conjugated double bonds and absence of oxygen atoms.

In the United States, purified chlorophyll is not an allowed color additive. However, juices from green vegetables are sometimes used as colorants in pasta (spinach pasta) and other foods.

Carotenoids form a large group containing hundreds of related compounds. They are present in photosynthesizing organisms, but their color is often masked by chlorophyll. This is most apparent in the fall of the year when chlorophyll in the leaves of deciduous trees breaks down revealing the brilliant reds, oranges, and yellows of the carotenoids that were there all along. Carotenoids may be divided into two groups: the *carotenes* that are hydrocarbons and the *xanthophylls* that contain oxygen in addition to hydrogen and carbon (Figures 15.5 and 15.6). Carotenoids are yellow, orange, or red and are insoluble in water. Most are fairly stable to heat and pH extremes but may be destroyed by oxidation. In intact tissue little oxidation takes place, but in processed foods where tissue has been disrupted, carotenoids may be oxidized by atmospheric oxygen. Oxidation rate may be affected by light, heat, and pro- or antioxidants.

A few carotenoids possess vitamin A activity meaning that they may be converted into retinol (vitamin A) in the body [5, 6]. Carotenoids with vitamin A activity include β-carotene, α-carotene, α-apo-8-carotenal, and cryptoxanthin [7].

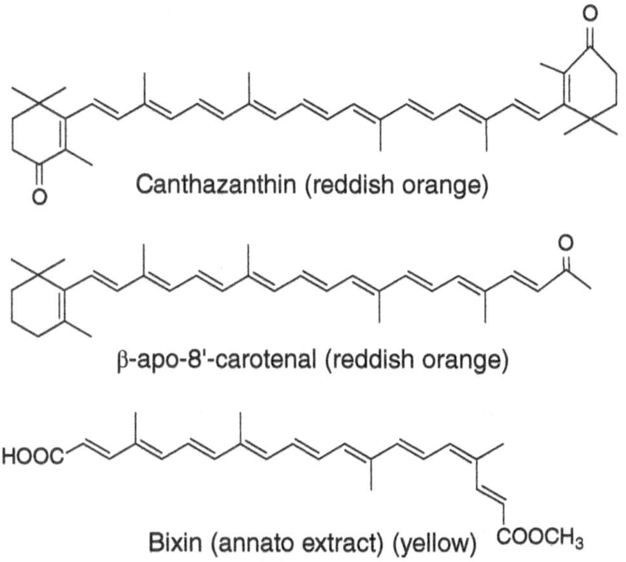

Canthazanthin (reddish orange)

β-apo-8'-carotenal (reddish orange)

Bixin (annato extract) (yellow) COOCH₃

Figure 15.6 Structures of selected xanthophylls. Note the presence of oxygen atoms in these structures.

Natural extracts and synthetic carotenoids are used to color margarine, butter, oils, beverages, soups, dairy and meat products, syrups, and macaroni. Bixin, a carotenoid present in annatto seeds, is widely used to produce yellow color in cheeses.

Three synthetic carotenoids currently approved by the FDA for food use are β-carotene (yellow to orange), β-apo-8-carotenal (orange to red), and canthaxanthin (red). Advantages of synthetic carotenoids include high purity, better control of concentrations and tolerances, and freedom from contaminating substances.

15.3 Apparatus and Instruments

1) Top loading balance
2) Blender
3) Test tubes
4) pH meter
5) Hot plate, or water bath at 95 °C
6) Beakers, 250 and 600 ml
7) Petri dish lid
8) Vortex mixer
9) Pipettors with tips
10) Filter and Whatman # 1 filter paper

15.4 Reagents and Materials

1) Raw and canned spinach; raw and cooked red cabbage; frozen green beans
2) Citrate/phosphate/borate/HCl buffer, pH: 2, 3, 5, and 7

3) Anhydrous sodium sulfate or anhydrous magnesium sulfate
4) Acetone
5) Silica gel TLC plates
6) Developing solvent: 60% petroleum ether/16% cyclohexane/10% ethyl acetate/10% acetone/4% methanol
7) Sand
8) Acetic acid, 0.1 N
9) Sodium bicarbonate, 0.1 M

15.5 Procedures

15.5.1 Extraction and Separation of Lipid Soluble Plant Pigments (Adapted from [8])

1) Weigh out 1.0 g of raw spinach.
2) Transfer to a mortar. Add 1 g anhydrous magnesium sulfate and 2 g sand.
3) Grind the mixture for 5–10 minutes until a light green powder is obtained.
4) Transfer to a test tube and add 4 ml acetone.
5) Mix well on a vortex mixer.
6) Let the mixture stand for 10 minutes.
7) Transfer 1 ml to a clean test tube.
8) On a 5 × 8 cm silica gel TLC plate, draw a line 1 cm from the bottom of the short edge **in pencil.** Using a pipettor, carefully spot, 1 cm apart, a small quantity of each extract on the line and label each lane.
9) Repeat Steps 1–8 using canned spinach. As long as the spots aren't too large, you may spot all treatments on one TLC plate.
10) Under the fume hood, put about 10 ml of developing solvent (60% petroleum ether/16% cyclohexane/10% ethyl acetate/10% acetone/4% methanol) into a 250 ml beaker and cover immediately with a petri dish lid (the solvent evaporates rapidly).
11) Place the spotted silica gel plate in the beaker, cover, and allow to develop until the leading edge of the solvent front is within 1–2 cm of the top of the plate, about 10 minutes.
12) Observe the bands in each lane.

15.5.2 Extraction of Water Soluble Plant Pigments

1) Weigh out 100 g of raw red cabbage.
2) Transfer to a blender jar. Add 200 ml distilled water. Blend for two minutes. Filter through Whatman # 1 filter paper. Save the filtrate.
3) Repeat steps 1–2 using red cabbage boiled in water for 10 minutes.
4) Compare the intensities and shades of colors between raw and cooked samples.

15.5.3 Effect of pH on the Color of Water Soluble Plant Pigments

1) Label large test tubes as follows: pH 2, 3, 5, and 7 (2 tubes for each pH). Add 5 ml of the appropriate buffer to each of the tubes.

2) Transfer 2 ml of your extracts prepared in Section 15.5.2 above to the tubes (do this so that you have each extract exposed to all 4 pH values.) Measure and record the pH of each buffer/extract mixture.

3) Observe the color of each solution and record your observations.

15.5.4 Demonstration

Add 100 ml 0.1 N acetic acid, 100 ml 0.1 M sodium bicarbonate, and 100 ml distilled water to separate 600 ml beakers. Bring to boil on hot plates. Add 50 g frozen green beans to each beaker. Bring back to boiling and boil an additional five minutes. Observe the color of the green beans from each treatment.

15.6 Study Questions

1 What were the principal pigments extracted from the spinach? The cabbage?

2 Did heat treatment affect the amount of pigment extracted?

3 Did pH affect the color of the cabbage extracts? Explain.

15.7 References

1 Lauro, G.J. (1991). A primer on natural colors. *Cereal Foods World* 36 (11): 949–953.
2 FDA (2020). Summary of color additives for use in the United States in foods, drugs, cosmetics, and medical devices. [cited 28 February 2020]. http://www.fda.gov/industry/color-additive-inventories/summary-color-additives-use-united-states-foods-drugs-cosmetics-and-medical-devices
3 Belitz, H.-D., Grosch, W., and Schieberle, P. (2009). *Food Chemistry*, 4e, 1070. Berlin: Springer.
4 Brady, J.W. (2013). *Introductory Food Chemistry*, 638. Ithaca: Comstock Publishing Associates.
5 Tang, G. (2010). Bioconversion of dietary provitamin A carotenoids to vitamin A in humans. *The American Journal of Clinical Nutrition* 91 (5): 1468S–1473S.
6 dela Seña, C., Riedl, K.M., Narayanasamy, S. et al. (2014). The human enzyme that converts dietary provitamin A carotenoids to vitamin A is a dioxygenase. *The Journal of Biological Chemistry* 289 (19): 13661–13666.
7 Bohn, T. and Provitamin, A. (2012). Carotenoids: occurrence, intake and bioavailability. In: *Vitamin A and Carotenoids: Chemistry, Analysis, Function and Effects* (ed. V.R. Preedy), 142–161. Cambridge: Royal Society of Chemistry. (Food and Nutritional Components in Focus).
8 Quach, H.T., Steeper, R.L., and Griffin, G.W. (2004). An improved method for the extraction and thin-layer chromatography of chlorophyll a and b from spinach. *Journal of Chemical Education* 81 (3): 385.

15.8 Suggested Reading

Frick, D. and Huck, P. (1995). Food color terminology. *Cereal Foods World* 40 (4): 209–218.

Schwartz, S.J., Cooperstone, J.L., Cichon, M.J. et al. (2017). Colorants. In: *Fennema's Food Chemistry*, 5e (eds. S. Damodaran and K.L. Parkin), 681–752. Boca Raton: CRC Press, Taylor & Francis Group.

Valverde, J., This, H., and Vignolle, M. (2007). Quantitative determination of photosynthetic pigments in green beans using thin-layer chromatography and a flatbed scanner as densitometer. *Journal of Chemical Education* 84 (9): 1505.

16

Meat Pigments

16.1   Learning Outcomes

After completing this exercise, students will be able to:

1) Prepare myoglobin and oxymyoglobin starting from a solution of metmyoglobin.
2) Record a visible spectrum of the various forms of myoglobin using a scanning spectrophotometer.
3) Use Beer's Law to calculate the concentrations of each of the three common forms of myoglobin in a solution.
4) Relate the chemistry of myoglobin to the variations in surface and interior colors of raw beef products stored under different conditions and for different times.
5) Explain why cooking changes the color of beef from red to brown.

16.2   Introduction

Color is an important quality attribute of fresh and processed meats. The major pigments in meats are myoglobin and hemoglobin. Myoglobin predominates in well-bled muscle tissue [1]. Myoglobin and hemoglobin are both globular heme-containing proteins. Myoglobin is monomeric (it contains a single polypeptide chain) and has a molecular weight of about 17,600 daltons. It is present in both skeletal and heart muscle and has the capacity to bind molecular oxygen within muscle. Hemoglobin is a tetramer with a molecular weight of about 64,000. The portion of both molecules that is responsible for meat color is the heme complex (Figure 16.1).

In fresh meat systems, myoglobin may exist either as myoglobin (sometimes called deoxymyoglobin), oxymyoglobin, or metmyoglobin. These three forms are important because they affect the color of the meat.

The iron atom at the center of the heme complex is key to myoglobin's function and color. When the oxidation state of the iron is +2 (ferrous), it is capable of binding molecular oxygen (O_2), nitric oxide (NO), carbon monoxide (CO), water, and other molecules. In myoglobin, the oxidation state of iron is +2. In oxymyoglobin, the oxidation state of iron is also +2 and molecular oxygen is bound to the iron. In metmyoglobin, the oxidation state of iron is +3 (ferric) and the ability to bind oxygen is lost. Myoglobin is a deep purple color; oxymyoglobin is red, and metmyoglobin is brown. Each of these forms has its own distinct spectrum (Figure 16.3).

Food Chemistry: A Laboratory Manual, Second Edition. Dennis D. Miller and C. K. Yeung.
© 2022 John Wiley & Sons, Inc. Published 2022 by John Wiley & Sons, Inc.
Companion website: www.wiley.com/go/Miller/foodchemistry2

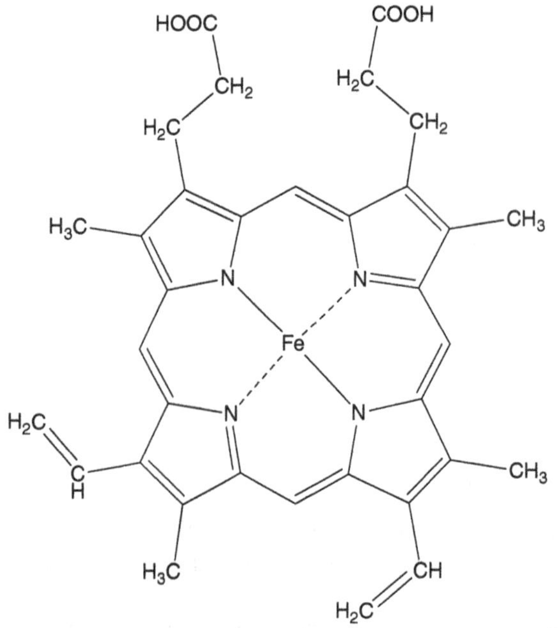

Figure 16.1 Heme. Heme is formed when an iron ion binds at the center of a protoporphyrin IX molecule. Iron is hexa-coordinate and therefore is capable of forming two additional coordinate covalent bonds. In oxymyoglobin and oxyhemoglobin, iron binds to a histidine residue in the globin chain and to molecular oxygen (see Figure 16.2).

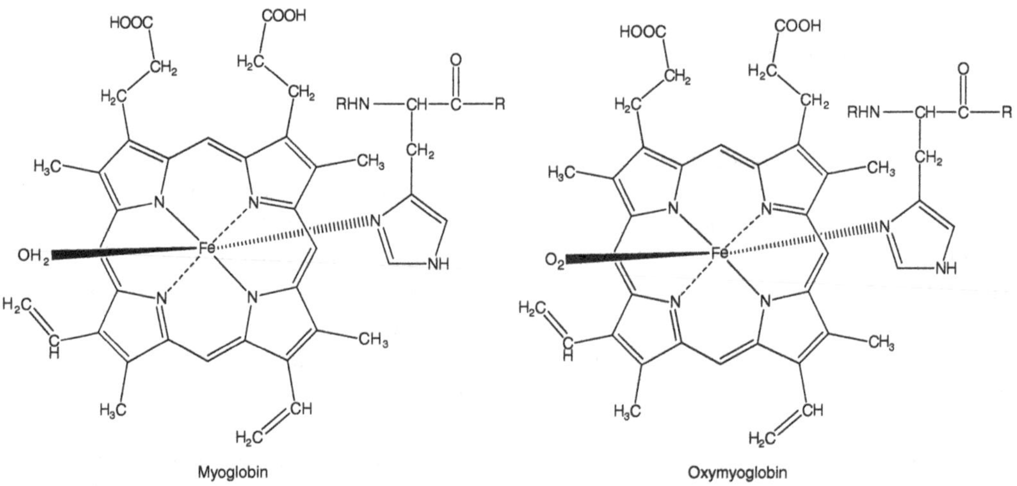

Myoglobin

Oxymyoglobin

Figure 16.2 Structures of myoglobin and oxymyoglobin showing iron binding to a histidine residue on the globin polypeptide chain and to water or oxygen in myoglobin and oxymyoglobin, respectively. The oxidation state of iron in myoglobin and oxymyoglobin is +2.

The chemical state of meat pigments can be controlled to some extent by regulating the partial pressure of oxygen in the meat environment. In fresh meat in the absence of atmospheric oxygen, myoglobin iron will be in the ferrous form. This is because normal enzyme activity in the meat uses up oxygen and provides a reducing environment [1]. If small amounts of oxygen are allowed to come into contact with the meat surface, the myoglobin will be oxidized to metmyoglobin. In the

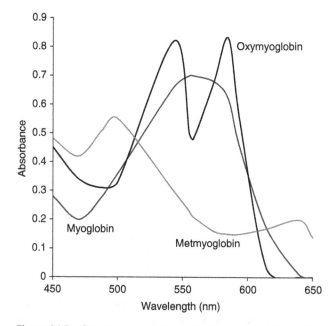

Figure 16.3 Spectral curves of myoglobin, oxymyoglobin, and metmyoglobin. Redrawn from [2].

presence of larger amounts of oxygen, the myoglobin will bind O_2 and thereby be converted to oxymyoglobin. In oxymyoglobin, the ferrous state is stabilized and is less likely to oxidize to the ferric form. Consumers prefer fresh meat that is red in color, especially fresh beef, and tend to pass over cuts that appear brown because color is often perceived by consumers as an indicator of meat quality [3].

Experimentally, Fe^{3+} in metmyoglobin can be reduced to Fe^{2+} by various reducing agents. A particularly effective reducing agent is sodium dithionite ($Na_2S_2O_4$), also known as sodium hydrosulfite. Half reactions for the oxidation of dithionite and the reduction of Fe^{3+} iron are shown below:

$$S_2O_4^{-2} + 4\,H_2O \longrightarrow 2\,HSO_4^- + 6\,H^+ + 6\,e-$$

$$6\,Fe^{+3} + 6\,e^- \longrightarrow 6\,Fe^{+2}$$

$$6\,Fe^{+3} + S_2O_4^{-2} + 4\,H_2O \longrightarrow 6\,Fe^{+2} + 2\,HSO_4^- + 6\,H^+$$

From these reactions, we can see how sodium dithionite could donate an electron to Fe^{3+}, reducing it to Fe^{2+}.

Ascorbic acid can also donate an electron to Fe^{3+}, reducing it to Fe^{2+}.

16.2.1 Meat Curing

Meat curing has been used for centuries as a preservation technique. Both nitrate and nitrite have been used in the meat curing process, but nitrate must be reduced to nitrite to be effective. A primary motivation for curing meat is the characteristic stable pinkish color that it produces. Curing meat with nitrite produces a unique flavor. It also inhibits lipid oxidation and retards the growth of spoilage and pathogenic bacteria [4].

When meat is cured, the myoglobin is converted to nitric oxide myoglobin (NOMb). This is accomplished by treating the meat with sodium nitrite and a suitable reducing agent to generate nitric oxide (NO) and to reduce heme Fe^{3+} to Fe^{2+} [1]. The chemistry of the production of NO from sodium nitrite is shown below (HRd represents a reducing agent) [5].

$$NaNO_2 + H^+ \longrightarrow HNO_2 + Na^+$$

$$2\,HNO_2 \rightleftharpoons N_2O_3 + H_2O$$

$$N_2O_3 + HRd \rightleftharpoons NO + Rd + HNO_2$$

Either ascorbic acid or erythorbic acid is commonly used as a reducing agent in the meat curing process.

16.2.2 Effect of Cooking on Meat Color

With the exception of nitrite cured meats, cooking causes meat color to turn brown. This change is due to the denaturation of the myoglobin and/or the oxidation of ferrohemochrome to ferrihemochrome (Figure 16.4). [6]. Many consumers and chefs rely on the color of cooked meat to determine doneness. The USDA recommends that beef steaks and roasts be cooked to an internal temperature of 145 °F and ground beef to 160 °F to ensure the inactivation of pathogenic organisms that may be present in the raw meat [7]. The recommended temperature is lower for intact cuts of meat (e.g. steaks and roasts) because pathogens generally do not penetrate into the interior of the meat. It is generally assumed that when the color of the innermost part of ground meat is no longer pink and the juices run clear, the meat is safe to eat. This assumption may be unwarranted because, under some conditions, the loss of pink color may occur before a safe temperature is

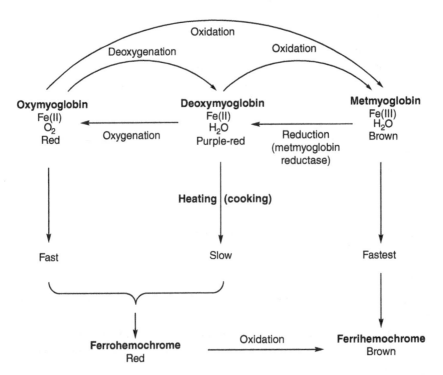

Figure 16.4 The three main forms of myoglobin pigments in meat, their relative stabilities to heat, and inter conversions between the forms [6].

reached due to the differences in heat sensitivity of the three common forms of myoglobin in meat (Figure 16.4). Deoxymyoglobin is the most stable to heat, so meats that contain high amounts of this form will not lose their pink color until a safe temperature is reached. In contrast, metmyoglobin denatures quickly and, moreover, metmyoglobin is brown even in raw meat. When hamburger is stored for long periods of time, the myoglobin in the interior of the patty will gradually oxidize to metmyoglobin so that the interior of the patty is no longer pink even when the meat is raw or the temperature is well below 160 °F [6]. Also, when meats are stored in high oxygen conditions to maintain their red color, a high proportion of the total myoglobin will be in the oxymyoglobin form. Oxymyoglobin is less stable than deoxymyoglobin and will denature at lower temperatures, causing the meat to turn brown at temperatures below the safe level [6]. Therefore, the USDA strongly recommends that consumers use a thermometer to determine doneness rather than relying of the color of the meat [7].

16.3 Apparatus and Instruments

1) Graduated cylinders, 10 and 50 ml
2) Test tubes and holders
3) Beakers, 100 ml
4) Parafilm
5) Water bath, 95 °C
6) Pipettes
7) Scanning spectrophotometer
8) Cuvettes
9) Knife and cutting boards
10) Vortex mixer

16.4 Reagents and Materials

1) Beef steaks packaged in either O_2 permeable film or O_2 impermeable film.
2) Buffer, pH 5.2 (23.2 mM citric acid + 53.4 mM disodium phosphate). De-aerated.
3) Buffer, pH 7.0 (1/15 M phosphate buffer). De-aerated.
4) Metmyoglobin (1.0 mg ml^{-1}) in pH 7 buffer. De-aerated after mixing.
5) Metmyoglobin (1.0 mg ml^{-1}) in pH 5.2 buffer. De-aerated after mixing.
6) Sodium dithionite ($Na_2S_2O_4$) crystals.
7) Sodium nitrite ($NaNO_2$), 1.0 M in water. De-aerated.
8) Ascorbic acid, 1.0 M in water. De-aerated.

16.5 Procedures

16.5.1 Preparation and Spectral Analysis of Myoglobin, Oxymyoglobin, and Metmyoglobin

1) Mark 4 test tubes at 10.0 ml, label them Tubes 1, 2, 3, and 4.
2) Obtain (in a 100 ml beaker) 50 ml of de-aerated metmyoglobin in pH 7 buffer. (The solutions were de-aerated by swirling them gently in a filter flask while pulling a mild vacuum with a

water aspirator. Be careful not to agitate the metmyoglobin solutions because agitation will increase O_2 uptake from the air.)

3) Carefully pour 10 ml of the metmyoglobin solution into each of the four tubes avoiding agitation as much as possible.
4) Treat each tube as indicated in the table:

Treatment name	Composition	Treatment
Tube 1: control	Metmyoglobin, $1\,mg\,ml^{-1}$ in phosphate buffer, pH 7	Gently invert tube 2x to mix
Tube 2: control + agitation	Metmyoglobin, $1\,mg\,ml^{-1}$ in phosphate buffer, pH 7	Vortex for 90s to aerate
Tube 3: metmyoglobin + reducing agent	Metmyoglobin, $1\,mg\,ml^{-1}$ in phosphate buffer, pH 7, + 10 mg sodium dithionite	Cap tube and gently invert 2x to mix
Tube 4: metmyoglobin + reducing agent + agitation	Metmyoglobin, $1\,mg\,ml^{-1}$ in phosphate buffer, pH 7, + 10 mg sodium dithionite	Vortex for 90s to aerate

5) Make careful observations of the color of each treatment.
6) Using a scanning spectrophotometer, record the absorption spectrum for each treatment by scanning between 475 and 650 nm.
7) Place the treated myoglobin solutions in a 95 °C water bath for 15 minutes. Describe changes that result from the heat treatment.

16.5.2 Preparation and Spectral Analysis of Nitric Oxide Myoglobin

1) Mark three test tubes at 10.0 ml and label them Tubes 5, 6, and 7.
2) Transfer 10 ml of the pH 5.2 metmyoglobin solution into the tubes.
3) Treat each tube as follows:

Treatment name	Composition	Treatment
Tube 5: control	10 ml metmyoglobin in pH 5.2 buffer + 2 ml water	Cap tube and invert gently to mix
Tube 6: metmyoglobin + $NaNO_2$	10 ml metmyoglobin in pH 5.2 buffer + 2 ml 1 M $NaNO_2$	Cap tube and invert gently to mix
Tube 7: metmyoglobin + $NaNO_2$ + ascorbic acid	10 ml metmyoglobin in pH 5.2 buffer + 1 ml 1 M $NaNO_2$ + 1 ml 1 M ascorbic acid	Cap tube and invert gently to mix

4) Repeat Steps 5 through 7 in Section 16.5.1.

16.5.3 Concentration of Metmyoglobin, Myoglobin, and Oxymyoglobin

Calculate the concentration in $mmol\,l^{-1}$ (mM) of the sum of all three myoglobin forms in each tube from Section 16.5.1 (assume that all the myoglobin is present in one of these forms). Recall that the concentration of the metmyoglobin solution that was prepared for you was $1\,mg\,ml^{-1}$ and that the molecular weight of metmyoglobin is $17,600\,mg\,mmol^{-1}$. Now calculate the concentration of the myoglobin in Tubes 1, 2, 3, and 4 using your absorbance readings and Beer' law. The millimolar extinction coefficient (ε) for metmyoglobin is $9.84\,\text{mM}^{-1}\text{cm}^{-1}$ at 503 nm; for myoglobin is $12.30\,\text{mM}^{-1}\text{cm}^{-1}$ at 557 nm; and for oxymyoglobin is $14.37\,\text{mM}^{-1}\,\text{cm}^{-1}$ at 582 nm. **Note**: these extinction coefficients were taken from Tang et al. [8].

Also recall the equation for Beer's law: $A = \varepsilon bc$ where A = absorbance, ε = the molar extinction coefficient ($\text{mM}^{-1}\,\text{cm}^{-1}$); b = light path length in the cuvette (1 cm), and c = concentration in mM.

The above calculations assume that our conversions between the various forms were 100% complete. Why might this assumption be invalid?

16.5.4 Demonstration

Fresh beef steaks were wrapped in films of differing oxygen permeability and stored for 48 hours at 4 °C. Observe the color of the various samples and explain any differences you see. Remove the films and observe for a few minutes. Note any color changes. Cut the steaks and observe the color of the interior of the meat.

16.6 Study Questions

1 Describe the changes that result from the heat treatment performed in Sections 16.5.1 and 16.5.2.

2 What purpose does the sodium dithionite serve? Is there another compound that could be used for the same purpose? If so, suggest one.

3 Explain any pigment differences you see in the steaks stored in the different films. What are they and why did they occur?

16.7 References

1 Aberle, E.D., Forrest, J.C., Gerrard, D.E., and Mills, E.W. (2012). *Principles of Meat Science*, 5e, 426. Dubuque, IA: Kendall Hunt Publishing.
2 Clydesdale, F.M. and Francis, F.S. (1976). Pigments. In: *Food Chemistry* (ed. O.R. Fennema), 385–426. New York: Marcel Dekker.
3 Font-i-Furnols, M. and Guerrero, L. (2014). Consumer preference, behavior and perception about meat and meat products: an overview. *Meat Science* 98 (3): 361–371.
4 Sindelar, J.J. and Milkowski, A.L. (2012). Human safety controversies surrounding nitrate and nitrite in the diet. *Nitric Oxide* 26 (4): 259–266.
5 Pegg, R.B., Shahidi, F., and Fox, J.B. Jr. (1997). Unraveling the chemical identity of meat pigments. *Critical Reviews in Food Science and Nutrition* 37 (6): 561–589.
6 King, J.W., Turner, N.J., and Whyte, R. (2006). Does it look cooked? A review of factors that influence cooked meat color. *Journal of Food Science* 71 (4): R31–R40.
7 USDA FSIS. Beef from farm to table [Internet]. [cited 2 March 2020]. https://www.fsis.usda.gov/wps/portal/fsis/topics/food-safety-education/get-answers/food-safety-fact-sheets/meat-preparation/beef-from-farm-to-table/ct_index

8 Tang, J., Faustman, C., and Hoagland, T.A. (2004). Krzywicki revisited: equations for spectrophotometric determination of myoglobin redox forms in aqueous meat extracts. *Journal of Food Science* 69 (9): C717–C720.

16.8 Suggested Reading

Bowen, W.J. (1949). The absorption spectra and extinction coefficients of myoglobin. *The Journal of Biological Chemistry* 179 (1): 235–245.

Eilert, S.J. (2005). New packaging technologies for the 21st century. *Meat Science* 71 (1): 122–127.

Fox, J.B. Jr. (1966). Chemistry of meat pigments. *Journal of Agricultural and Food Chemistry* 14 (3): 207–210.

Honikel, K.-O. (2008). The use and control of nitrate and nitrite for the processing of meat products. *Meat Science* 78 (1): 68–76.

Livingston, D.J. and Brown, W.D. (1981). The chemistry of myoglobin and its reactions. *Food Technology* 35 (5): 244–252.

Mancini, R.A. and Ramanathan, R. (2020). Molecular basis of meat color. In: *Meat Quality Analysis* (eds. A.K. Biswas and P.K. Mandal), 117–129. Cambridge, MA: Academic Press.

Millar, S.J., Moss, B.W., and Stevenson, M.H. (1996). Some observations on the absorption spectra of various myoglobin derivatives found in meat. *Meat Science* 42 (3): 277–288.

Schwartz, S.J., Cooperstone, J.L., Cichon, M.J. et al. (2017). Colorants. In: *Fennema's Food Chemistry*, 5e (eds. S. Damodaran and K.L. Parkin), 681–752. Boca Raton: CRC Press, Taylor & Francis Group.

Suman, S.P. and Poulson, J. (2013). Myoglobin chemistry and meat color. *Annual Review of Food Science and Technology* 4: 79–99.

17

Meat Tenderizers

17.1 Learning Outcomes

After completing this exercise, students will be able to:

1) Describe the principles underlying the technique of electrophoresis.
2) Apply electrophoresis to assess the action of meat tenderizing enzymes.
3) Understand and compare the actions of two meat tenderizing enzymes.
4) Explain the effects of meat tenderizers on marshmallows.

17.2 Introduction

Tenderness is an important quality factor in meats. The perceived tenderness of a meat sample is influenced by a complex set of factors with muscle proteins playing a major role. Three classes of protein appear to be involved in tenderness [1]. They are connective tissue proteins (collagen and elastin), contractile (myofibrillar) proteins (actin, myosin, tropomyosin, titin, and nebulin), and sarcoplasmic proteins (glycolytic and other enzymes and myoglobin). These proteins may be classified according to their solubility in various solvents [1]. Connective tissue proteins are insoluble in water and in concentrated salt solutions. Contractile proteins are insoluble in water and dilute salt solutions but are soluble in concentrated salt or urea solutions. The sarcoplasmic proteins are soluble in water or dilute salt solutions. Presumably, these proteins impart toughness to meat because they interact and cross link to produce large molecular weight structures that are difficult to break apart by chewing.

A common strategy for tenderizing meat is to marinate in an acidic marinade prior to cooking. Many marinades contain vinegar along with other ingredients such as vegetable oil, soy sauce, spices, and so on. The acetic acid in the vinegar promotes acid hydrolysis of proteins in the meat, thereby producing a tenderizing effect.

Proteolytic enzymes capable of partially hydrolyzing muscle proteins may also be used to tenderize meat. One strategy to promote hydrolysis of muscle proteins is to "condition" the meat. Conditioning involves holding carcasses at chill temperatures for 10–14 days before cutting. Conditioning is an effective means for enhancing tenderness. Presumably the improved tenderness is due to the action of endogenous proteolytic enzymes such as cathepsin and calpain [1, 2].

Food Chemistry: A Laboratory Manual, Second Edition. Dennis D. Miller and C. K. Yeung.
© 2022 John Wiley & Sons, Inc. Published 2022 by John Wiley & Sons, Inc.
Companion website: www.wiley.com/go/Miller/foodchemistry2

Tenderizing meat artificially by adding exogenous proteolytic enzymes has been in practice for hundreds of years. Indigenous Peoples of Mexico are known to have wrapped meat in papaw leaves during cooking [1]. Papaw leaves contain a proteolytic enzyme called papain. Today, several enzymes are used commercially and in the home to tenderize meat. They include the plant-derived enzymes ficin (from figs), papain (from papaw leaves), and bromelain (from pineapple). These enzymes may be added to the meat post-slaughter or injected into the blood stream of the live animal 1–30 minutes before slaughter [1]. Injecting into the live animal is preferred because it allows for a more even distribution of the enzyme in the muscles.

The use of these enzymes is not without problems. As proteolytic enzymes go, they are relatively nonspecific and are capable of hydrolyzing many different proteins. This can lead to over tenderization when hydrolysis is too extensive [3]. Treatments that disrupt connective tissue (collagen and elastin) while minimizing hydrolysis of myofibrillar proteins are preferred [4]. It has been suggested, therefore, that enzymes that are specific for collagen and elastin would be preferable to ficin, papain, and bromelain [3]. Takagi et al. [3] have shown that an elastase produced by a strain of *Bacillus* has high hydrolyzing activity toward collagen and elastin but low activity toward myofibrillar proteins. They suggest that this enzyme may be preferable to the commonly used plant-derived enzymes for meat tenderization.

In order to understand and monitor the actions of meat tenderizing enzymes, food scientists have relied on a powerful technique called electrophoresis. In this technique, proteins are solubilized by extraction with appropriate solvents, applied to a gel, and placed in an apparatus designed to expose the proteins to an electric field. The electric field causes the charged protein molecules migrate through the gel. The rate of migration will be a function of the size of the protein allowing separation of the mixture of proteins in the sample. The molecular weight (MW) of unknown proteins may be determined by comparing their migration rates with standard proteins of known MW since proteins of similar MW will travel at the same rate and migrate the same distance within the gel. See Appendix for a detailed explanation of the principles and practice of electrophoresis.

We would expect proteolytic enzymes to reduce the MW of proteins by cutting them up into smaller peptides. Thus, electrophoresis should be an ideal technique for studying the action of meat tenderizers. In this experiment we will use electrophoresis to study the action of two meat tenderizers, papain and bromelain, on lean ground beef.

A simpler approach to observing the proteolytic action of plant-based enzymes is to apply fresh fruit juices to marshmallows. The primary ingredients in marshmallows are sugar (usually a combination of granulated sugar, confectioners' sugar, and corn syrup) and unflavored gelatin [5]. Gelatin is a protein made from collagen. To make marshmallows, the sugars and corn syrup are mixed with water and heated to around 115 °C. The hot syrup is mixed into a gelatin solution and whipped to a stiff foam. The resulting marshmallows are basically soluble sugars and dextrins in a protein matrix. The protein matrix is insoluble in water but will dissolve when the proteins are hydrolyzed. To perform a qualitative assay of protease activity, one can simply add solutions of proteolytic enzymes to marshmallows and see if they dissolve.

17.3 Apparatus and Instruments

1) Balance
2) Test tubes and holders
3) Water bath, boiling

4) Bench-top centrifuge
5) Small centrifuge tubes
6) Pipettor and Bio-Rad Prot/elec tips
7) Pre-cast mini protean mini gel (7.5% single percentage polyacrylamide gel)
8) Power supply
9) Apparatus for running mini gels
10) Trays for staining and destaining gels
11) Small beakers and watch glasses

17.4 Reagents and Materials

1) Lean ground beef (ground round steak)
2) Papain (1 mg ml^{-1} in water) (2 x crystallized, 80% protein; Sigma Chemical Co., St Louis)
3) Bromelain (1 mg ml^{-1} in water) (50% protein, Sigma Chemical Co., St. Louis)
4) 6 M urea solution containing 2% sodium dodecyl sulfate (SDS)
5) Sample buffer: 100 mM Tris-HCl containing 4% SDS, 20% glycerol, 0.1% bromphenol blue, and 5 % mercaptoethanol.
6) Running buffer: Bio-Rad 10X Tris-glycine SDS buffer diluted 1:10.
7) Coomassie Blue staining solution: 0.25% Coomassie blue R in methanol:acetic acid:water, 45:10:45.
8) Destaining solution: methanol:acetic acid:water, 45:10:45.
9) Bio-Rad SDS-PAGE Standards; broad range: 6.5–200 kdaltons
10) Mini marshmallows
11) Cubes of fresh, uncooked beef steak
12) White vinegar
13) Fresh pineapple juice
14) Canned pineapple juice

17.5 Procedures

The following procedure is adapted from the method of Kim and Taub [6].

17.5.1 Preparation of Samples and Standards

17.5.1.1 Sample Treatments

1) Transfer 0.8 g of lean ground beef to each of six test tubes.
2) Add 0.8 ml papain solution to three of the tubes and 0.8 ml bromelain solution to the other three. Mix thoroughly.
3) Immediately place two of the tubes (one papain and one bromelain) in a boiling water bath for two minutes to inactivate the enzymes (0 time treatment).
4) Incubate two of the tubes (one papain and one bromelain) for 30 minutes at room temperature (30 minutes treatment). Inactivate the enzymes in a boiling water bath.
5) Incubate the other two tubes for 60 minutes at room temperature (60 minutes treatment). Inactivate the enzymes in a boiling water bath.

17.5.1.2   Protein Extraction and Preparation for Electrophoresis

1) Add 5 ml 6 M urea containing 2% SDS to each of the tubes in Section 17.5.1.1 above. Mix thoroughly and allow to stand for 10 minutes. Mix again.
2) Centrifuge at moderate speed in a bench-top centrifuge.
3) Mix 1 ml of each supernatant with 1 ml of sample buffer. Mix thoroughly.

17.5.1.3   Preparation of SDS-PAGE Standards for Electrophoresis.

1) Mix 1 ml of the standard solution with 1 ml of sample buffer.

17.5.2   Electrophoresis

17.5.2.1   Loading and Running the Gel

1) Assemble the electrophoresis apparatus and fill with the running buffer.
2) Pipet 10 μl of the SDS-PAGE standard into a sample well in the middle of the gel. (Use a Bio-Rad Prot/elec tip on a pipettor to load the samples on the gel.)
3) Pipet 20 μl of each sample into the other sample wells. Be certain to record the location of each sample and standard in your lab notebook.
4) Run the gels at 200 volts for approximately 30 minutes or until the tracking dye (bromphenol blue) reaches the bottom of the gel.

17.5.2.2   Staining the Gel

1) Remove gels from the apparatus and stain for two hours in Coomassie blue staining solution.
2) Remove the staining solution and add destaining solution. After two hours, pour off the destaining solution and add fresh destaining solution. Leave in the fresh destaining solution overnight.
3) Observe gels. The proteins will appear as blue bands on a clear background.

17.5.3   Demonstration

Place a mini marshmallow in each of six small beakers. Cover the marshmallows with 25 ml of the following liquids: water, papain solution (1 mg ml^{-1}), bromelain solution (1 mg ml^{-1}), fresh pineapple juice, canned pineapple juice, white vinegar. Cover the beakers with watch glasses and leave to sit overnight at room temperature. Observe the marshmallows and explain the differences between the various treatments.

Repeat with 1 cm cubes of fresh, uncooked beef steak.

17.6   Study Questions

1   Compare the intensity and location of the protein bands on your gels. Explain any differences between incubation times and enzymes.

2   The molecular weight of myosin is about 200 kdaltons. Is there a protein in your samples with this molecular weight?

3   Actin has a molecular weight of about 43–48 kdaltons. Is there a protein in your samples with this molecular weight?

4   Compare and explain the differences between the various treatments in the demonstration.

17.7 References

1 Lawrie, R.A. (1991). *Meat science*, 5e, 293. Oxford, England; New York, USA: Pergamon Press.
2 Huff-Lonergan, E., Mitsuhashi, T., Beekman, D.D. et al. (1996). Proteolysis of specific muscle structural proteins by µ-calpain at low pH and temperature is similar to degradation in postmortem bovine muscle. *Journal of Animal Science* 74 (5): 993–1008.
3 Hiroshi, T., Masaaki, K., Tomoaki, H. et al. (1992). Effects of an alkaline elastase from an alkalophilic Bacillus strain on the tenderization of beef meat. *Journal of Agricultural and Food Chemistry* 40 (12): 2364–2368.
4 Tantamacharik, T., Carne, A., Agyei, D. et al. (2018). Use of plant proteolytic enzymes for meat processing. In: *Biotechnological Applications of Plant Proteolytic Enzymes* (eds. M.G. Guevara and G.R. Daleo), 43–67. Cham: Springer International Publishing.
5 Hartel, R.W. and Hartel, A. (2014). *Candy Bites: The Science of Sweets*, 269. New York, NY: Copernicus Books.
6 Kim, H.-J. and Taub, I.A. (1991). Specific degradation of myosin in meat by bromelain. *Food Chemistry* 40 (3): 337–343.

17.8 Suggested Reading

Ha, M., Bekhit, A.E.-D., Carne, A., and Hopkins, D.L. (2013). Comparison of the proteolytic activities of new commercially available bacterial and fungal proteases toward meat proteins. *Journal of Food Science* 78 (2): C170–C177.
Hagar, W.G. and Bullerwell, L.D. Supermarket proteases [Internet]. NSTA News. [cited 28 February 2020]. https://www.nsta.org/publications/news/story.aspx?id=48656

18

Detection of Genetically Engineered Maize Varieties

18.1 Learning Outcomes

After completing this exercise, students will be able to:

1) Explain why glyphosate is toxic to most plants.
2) Describe the principles underlying PCR.
3) Describe the principles underlying agarose gel electrophoresis.
4) Describe the principles underlying lateral flow technology methods.
5) Determine whether a food sample contains an ingredient from a genetically engineered crop using two different methodologies.

18.2 Introduction

Genetically engineered (GE) seeds, commonly known as GMOs, first became widely available to farmers in the mid-1990s. These seeds were enthusiastically received by farmers due to the beneficial agronomic traits they offer. These traits include insect resistance (e.g. Bt corn and Bt cotton), herbicide resistance (HR) (e.g. Roundup Ready® corn, soybean, cotton, canola, and sugar beet), and virus resistance (e.g. virus resistant papaya and eggplant). The rapid rate of adoption of GE corn, cotton, and soybeans by US farmers is shown in Figure 18.1. Today, more than 75% of planted acres of these crops are planted with genetically engineered seeds. Organic foods presumably do not contain GE ingredients because USDA guidelines for organically grown crops prohibit the use of GE varieties [1].

Insect-resistant crops contain a gene that encodes for a protein that is toxic to certain insects but not to mammals. When these insects begin chewing on the GE plant, they quickly die from exposure to the toxic protein. The gene is present naturally in the soil bacterium *Bacillus thuringiensis* (Bt). The gene is inserted into the crop genome using recombinant DNA technology.

Herbicide tolerant crops are resistant to the herbicide glyphosate. Glyphosate is an effective herbicide because it is a potent inhibitor of the enzyme 5-enolpyruvyl-shikimate-3-phosphate synthase (EPSPS). EPSPS is a key enzyme in the shikimate pathway for the synthesis of aromatic amino acids (phenylalanine, tyrosine, and tryptophan) in plants. It catalyzes the addition of

Food Chemistry: A Laboratory Manual, Second Edition. Dennis D. Miller and C. K. Yeung.
© 2022 John Wiley & Sons, Inc. Published 2022 by John Wiley & Sons, Inc.
Companion website: www.wiley.com/go/Miller/foodchemistry2

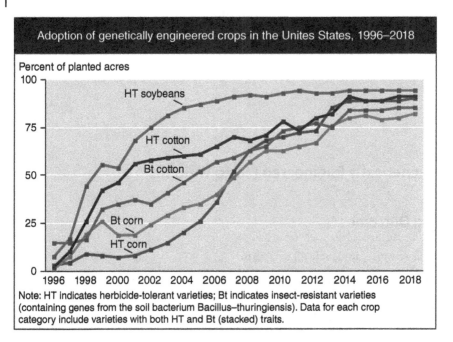

Figure 18.1 Trends in the adoption of genetically engineered crops by farmers in the United States [2].

phosphoenol pyruvate to shikimate-3-phosphate (Figure 18.2). Glyphosate, the active ingredient in Roundup®, is a widely used herbicide. As shown in Figure 18.3, the structure of glyphosate is similar to that of phosphoenol pyruvate. Glyphosate binds irreversibly to the active site of EPSPS, thereby blocking the synthesis of aromatic amino acids in the plant [3]. Without these amino acids, the plants cannot synthesize proteins and therefore they die. Since all plants rely on EPSPS for the synthesis of aromatic amino acids, glyphosate will kill virtually all plants it contacts. This makes it very effective for killing weeds, but it also kills crop plants and that limited its use until glyphosate-resistant (GR) varieties were developed.

The discovery of a glyphosate-resistant EPSPS in *Agrobacterium sp.* strain CP4 lead to the development of transgenic crops that are resistant to glyphosate [4]. This made it possible to control weeds in fields where these GR crops were growing by spraying them with glyphosate. Presumably, CP4 EPSPS is resistant to glyphosate because of an amino acid substitution in the active site that prevents the binding of glyphosate by CP4-EPSPS but does not prevent the catalytic activity of the enzyme [5].

While these GE seeds were welcomed and rapidly adopted by farmers, several activist groups began to raise questions about the safety of foods made from these crops and the possibility of environmental damage caused by the crops themselves or by glyphosate. This has led to a growing concern among consumers about these crops and campaigns to require food companies to label foods as GE foods. As a result, it became important to be able to distinguish between foods that contain genetically engineered ingredients from those that do not. This can be achieved either by testing for the presence of the gene that was inserted into the genome of the plant or by testing for the specific protein that the inserted DNA encodes. Polymerase chain reaction (PCR) methods are used to detect genes. Immunoassays are widely used to detect specific proteins in complex matrices.

In this experiment, we will test samples of corn meal or corn flour for the Roundup Ready® gene using two methods, PCR and immunoassay.

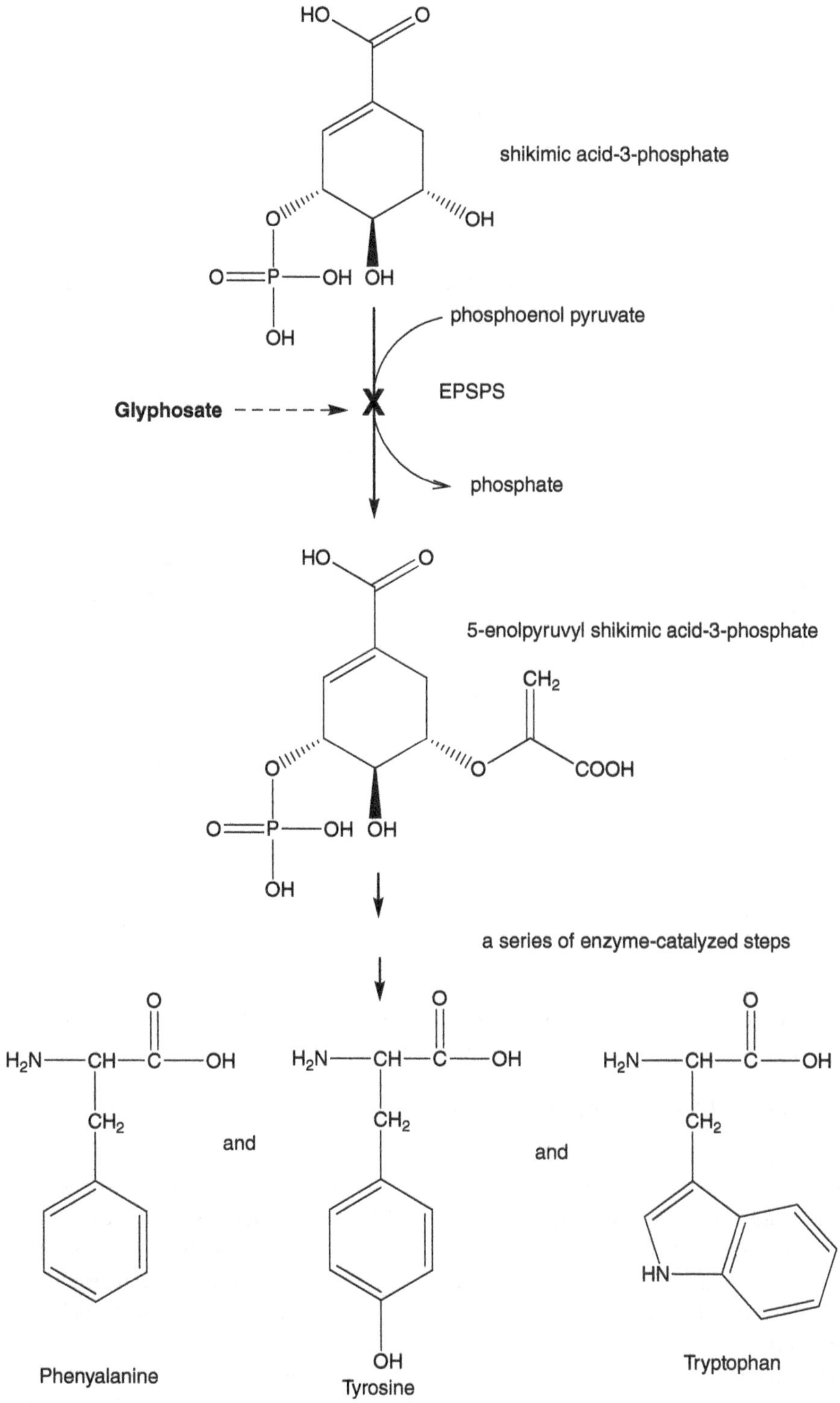

Figure 18.2 Steps in the shikimate pathway for the synthesis of aromatic amino acids in plants. Glyphosate binds irreversibly to EPSPS blocking the conversion of shikimic acid-3-phosphate to 5-enolpyruvyl shikimic acid-3-phosphate. Without 5-enolpyruvyl shikimic acid-3-phosphate, the plant cannot synthesize aromatic amino acids. Roundup Ready® plants contain a trans gene that encodes for a glyphosate-tolerant EPSPS, which makes them resistant to glyphosate. Adapted from [4].

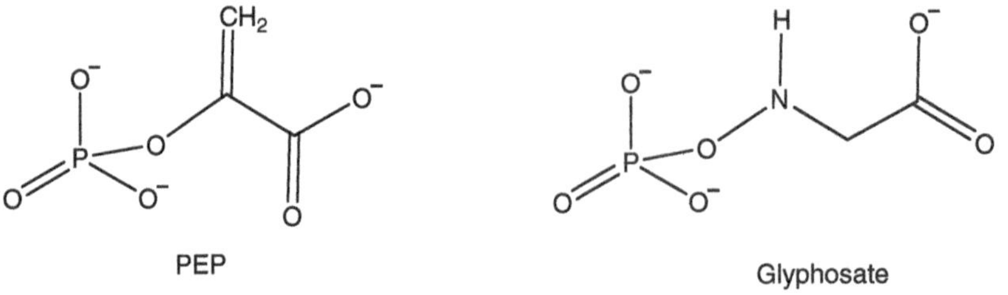

PEP

Glyphosate

Figure 18.3 Structures of phosphoenol pyruvate (PEP) and glyphosate. Notice that both structures contain a phosphate group and a carboxylate group, making them similar.

Lateral flow technology

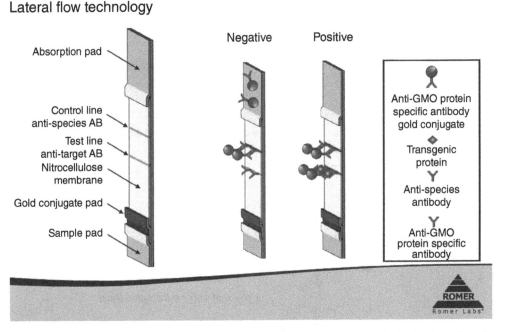

Figure 18.4 Diagram showing the structure of a typical lateral flow device (LFD) for the detection of a GMO protein. Reproduced from a slide set produced by Romer Labs®.

18.2.1 Detection of a GE Protein by Immunoassay

Immunoassays involve the use of antibodies that are specific for a given antigen, e.g. a transgenic protein in a genetically engineered plant. Several companies have used this technology to develop assays that can rapidly identify GMO crops in the field. The most widely used of these are the so-called lateral flow devices (LFD), also known as test strips. These devices may be used for either qualitative (a yes or no answer) or semi-quantitative determinations. LFDs are constructed by applying a strip of nitrocellulose membrane to a plastic backing and adding antibodies specific for the protein of interest (Figure 18.4).

A sample pad is applied to one end of the strip. This pad will be dipped in the protein-containing extract of the sample. Just above the sample pad is another pad that contains the antibody that is specific for the GE protein antigen of interest. This antibody is labeled with a colored marker, usually colloidal gold, and is mobile, i.e. it will move up the strip with the extract. The device contains another antibody specific to the GE protein, but this time it is immobilized to the strip upstream from the mobile conjugate pad and is not labeled with the colored marker. This is called the capture antibody. A second capture antibody is immobilized in a band above the first capture antibody. This antibody is specific for the mobile antibody, not the GE protein. See the paper by Grothaus et al. [6] for a detailed explanation of how this assay works.

To do the test, the sample, e.g. a corn kernel, is ground, and the protein is extracted. The bottom of the test strip is submerged in the sample extract. The extract solution moves upward by capillary action. As it passes through the pad containing the gold-labeled antibody, the GE protein binds to the antibody and the complex moves upward along with excess gold-labeled antibody not complexed with the GE protein. When the solution reaches the band containing the capture antibody, the GE protein-gold-labeled antibody complex is trapped and, as it accumulates, a red-colored band becomes visible (note that nanoparticles of gold appear red). Excess gold-labeled antibody continues moving up the strip until it reaches the band containing the immobilized antibody against the gold-labeled antibody where it accumulates in a second red band. The purpose of the second band is to confirm that the assay is working, i.e. that the gold-labeled antibody is moving up the strip. If a colored band appears below the control line, the test is positive for the GE protein. If there is no band below the control line, the test is negative. If no bands appear, the test is invalid and should be repeated.

Note that heating may denature proteins. This often makes soluble proteins insoluble. Insoluble proteins will not be extracted and therefore will not be present in the solution moving up the lateral flow device.

18.2.2 Detection of a Trans Gene by PCR

Polymerase chain reaction (PCR) methods are widely used in a range of fields for the identification of specific genes. For example, PCR is used in forensic science to identify crime suspects, in food science to detect food fraud, in food microbiology to identify foodborne pathogens, and in many other applications. In this experiment, we will use PCR to determine whether samples of corn meal contain a genetically engineered transgene that confers resistance to the herbicide glyphosate.

Students should review the basic principles of PCR methodology that they learned in previous courses. The reference by Namuth [7] provides a good review.

Basically, PCR methods are used to amplify a DNA fragment of known sequence from an organism of interest by using a DNA polymerase enzyme (this has been called molecular photocopying). Amplification produces billions of copies of a selected fragment of a gene in a short period of time, making it possible to identify the presence of the gene in an organism from a small amount of extracted DNA.

One big advantage of PCR is that all necessary components of the reaction can be combined in a single tube. These components include the following:

1) Template DNA (this is the DNA that you extract from the sample you are analyzing).
2) Primers (a forward primer and a reverse primer). Primers are short segments of DNA of known sequence. They are used to identify the region on the template you wish to copy. Sequences of primers for detecting the CP4-EPSPS gene have been published [5] and are shown below. They are available for purchase from commercial biotechnology supply companies.
   a) (F): 5'-ATGAATGACCTCGAGTAAGCTTGTTAA-3'
   b) (R): 5'-AAGAGATAACAGGATCCACTCAAACACT-3'
3) DNA polymerase. The most commonly used DNA polymerase is *Taq* polymerase. Unlike most enzymes, it is not denatured at temperatures near 100 °C, which is important since high temperatures are required to denature the template DNA (see below). *Taq* polymerase was originally isolated from a thermophilic bacterium called *Thermus aquaticus*. *T. aquaticus* is able to

grow in hot springs and that is where it was originally discovered. The optimal temperature for *Taq* polymerase activity is around 75 °C, but the enzyme is stable at temperatures approaching the boiling point of water.

4) Deoxyribonucleotide phosphates (dNTPs): adenine (A), guanine (G), cytosine (C), and thymine (T). A, G, C, and T are required for DNA replication.
5) Buffer to maintain a pH that is optimal for *Taq* polymerase activity.
6) Water. Molecular biology grade water certified to be free of DNase and RNase is recommended.

The first step in a PCR analysis is extraction of DNA from the sample. This can be tricky with plant samples since they contain substances such as polysaccharides and polyphenolics that can inhibit the PCR. In addition, since plant cells have protective cell walls, it can be difficult to lyse the cells. Kits are available from commercial companies that contain reagents that facilitate cell lysis and enable the removal of PCR inhibitors. One such kit is the DNeasy Plant Pro Kit by QIAGEN [8].

The second step is amplification of a fragment of the gene that encodes for the trait of interest. The amplification process involves three steps:

1) Denaturation of the DNA in the sample. This is accomplished by heating the extracted DNA to a temperature high enough to break the hydrogen bonds that hold the strands of double-stranded DNA together. Normally, this requires temperature in the 95–100 °C range. The result is single-stranded DNA.
2) Primer annealing. In this step, the primers bind to regions of the single-stranded DNA that contain sequences that are complementary to the sequence of the primers. The temperature for this step is lower to allow the primers to hydrogen bond to the single-stranded DNA in the sample. Normally, temperatures in the range of 45–55 °C are used for this step.
3) DNA extension. In this step, which is carried out at 72 °C, the DNA polymerase enzyme extends the primer in the 5' → 3' direction using the segment of the sample DNA as the template. The reaction continues until the DNA strand is extended to the end of the template region to be copied, producing a new DNA strand that is complementary to the template DNA sequence.

These reactions are conducted in a thermocycler, which is an instrument that rapidly cycles the temperature between the three temperatures selected. Once a cycle of the reaction is complete, the three-step process is repeated over and over, usually 25–35 times. Each cycle doubles the number of copies of the DNA fragment, so the amplification proceeds as an exponential rate. After 30 cycles, there will be 10^9 copies of the targeted DNA fragment! In our case, the fragment will contain 108 base pairs (bp). This is important to know because we will identify the fragment based on its size.

The final step is to analyze the PCR products using agarose gel electrophoresis to determine whether our sample contains a 108 base pair DNA fragment. If it does, our sample will be positive for the CPR-EPSPS gene. If it does not contain a DNA fragment of this size, we will conclude that our sample is not glyphosate tolerant.

Agarose gel electrophoresis is a very effective method for separating DNA fragments by size and visualizing them with a dye that binds to DNA. Agarose is a polysaccharide isolated from red algae seaweeds. At low concentrations in water, it forms a gel. A gel is a three-dimensional network that forms when the individual polymers associate, usually via non-covalent linkages. Gels are porous, and the pore size is inversely proportional to the concentration of the polymer (agarose in this case) in the gel. DNA molecules are negatively charged at neutral pH due to the phosphate groups in the DNA backbone. Therefore, when an electric field is applied to a gel containing DNA, it will migrate in the direction of the anode (which is positively charged). The charge density (mass/charge ratio) of DNA is uniform regardless of the size of the molecule. Therefore, in agarose gel electrophoresis, the rate at which the DNA moves through the gel is inversely proportional to the size of the DNA fragment, i.e. small fragments will migrate faster, large fragments more slowly. Agarose gel electrophoresis can separate DNA fragments ranging in size from around 100 bp to 25,000 (25 kb) bp [9]. When the electrophoresis run is complete, DNA fragments will accumulate in bands at varying distances from the origin. The DNA can be visualized by staining with a dye that is visible under either UV or visible light, depending on the dye used. The most common dye used in DNA electrophoresis is ethidium bromide. Unfortunately, ethidium bromide is a suspected carcinogen so we recommend an alternative dye if possible. One such dye is GelRed®, which is available through VWR (Radnor, PA) and other suppliers of molecular biology reagents.

18.3 Apparatus and Instruments

1) Balances
2) Micropipettes
3) Pipette tips
4) Microwave or hot plate
5) PCR thermal cycler
6) PCR tubes
7) Electrophoresis apparatus

18.4 Reagents and Materials

1) Corn meal or corn flour samples from Roundup Ready® and organically grown varieties.
2) Lateral Flow Immunoassay Kit for Detecting CP4 EPSPS in corn. Romer Labs AgraStrip® RUR Bulk Grain Strip Test – Qualitative Method [10] is recommended but other kits will also work well.
3) DNeasy Plant Pro Kit by QIAGEN [8] or a similar kit for DNA extraction.
4) Master mix for PCR amplification (AmpliTaq Gold® 360 PCR Master Mix [Thermo Fisher Scientific, Waltham, MA] or similar master mix).
5) Forward and reverse primers for CP4 EPSPS gene ([5], amplicon is 108 bp):
   a) (F): 5'-ATGAATGACCTCGAGTAAGCTTGTTAA-3'
   b) (R): 5'-AAGAGATAACAGGATCCACTCAAACACT-3'
   Note: Others have used different primers for the EPSPS gene to yield an amplicon of 256 bp [11]. You may want to try both to see which works best:
   c) (F): ACCGGCCTCATCCTGACGCT
   d) (R): CCGAGAGGCGGTCGCTTTCC
6) Master mix with added primers (this will be prepared by the teaching assistant). Amount for one PCR: 25 μl AmpliTaq Gold® 360 PCR Master Mix + 5 μl of $1\,\mu mol\,l^{-1}$ forward primer, 5 μl of $1\,\mu mol\,l^{-1}$ reverse primer + 5 μl water. This should be prepared just before use.
7) Agarose.
8) TAE buffer: $40\,mmol\,l^{-1}$ Tris, $20\,mmol\,l^{-1}$ acetic acid, $1\,mmol\,l^{-1}$ EDTA.

9) Gel loading solution: Pre-prepared solutions are commercially available. They may contain sucrose or glycerol plus a tracking dye. The sucrose or glycerol increases the density of the solution allowing the sample to settle to the bottom of the well in the gel. Bromophenol blue or amaranth are common tracking dyes. Tracking dyes should migrate ahead of the DNA fragments during the electrophoresis run, serving as an indicator for when to stop the run. We recommend amaranth for this experiment since it is small enough to run faster than a 100 bp DNA fragment.
10) GelRed® Nucleic Acid Gel Stain, 10,000 X or similar product
11) 100 bp DNA Ladder (from Invitrogen™ or similar)

18.5 Procedures

Obtain two samples (A and B) of corn meal or corn flour for analysis. These samples may or may not contain the CP4 EPSPS gene. It is your job to determine whether or not they do.

1) Immuno Assay for CP4 EPSPS (based on protocol described in package insert for the AgraStrip® RUR Bulk Grain Strip Test – Qualitative Method from Romer Labs® [10]). Kits supplied by other companies will also work. The following protocol is general and specifics may differ depending on the kit. It is best to follow the directions supplied by the manufacturer of the kit.
   a) Place 20 g of each corn meal sample in separate tubes.
   b) Add 25 ml distilled water and shake to mix.
   c) Allow to settle and transfer 500 µl of the supernatants to clean microcentrifuge tubes.
   d) Insert the test strip into the sample so that the liquid covers the sample pad but does not reach the conjugate pad.
   e) Leave the strip in the tube for five minutes.
   f) Remove the strip, photograph it, and ascertain whether the sample is genetically engineered or not.
2) Polymerase Chain Reaction
   a) Extract DNA from each sample using the DNeasy Plant Pro Kit by QIAGEN [8] or a similar kit.
   b) Transfer 40 µl master mix to a PCR tube.
   c) Add DNA extract. Amount will depend on the efficiency of the extraction. 1–10 µl is usually sufficient.
   d) Program PCR thermal cycler for five minutes at 95 °C to denature DNA; 30 cycles at 95 °C for 30 seconds, 50 °C for 30 seconds, and 72 °C for 30 seconds, and final extension at 72 °C for seven minutes.
   e) Place tubes in thermal cycler and run.
   f) Store tubes at −20 °C until you are ready to run the electrophoresis.
3) Agarose Gel Electrophoresis. The following protocol is adapted from Swope et al. [5] and Lee et al. [9]
   a) Gel preparation (most electrophoresis gel trays require 100 ml of agarose solution).
      i) Make a 2% (wt/vol) solution of agarose in the TAE buffer. To do this, transfer 4 g agarose to a 500 ml Erlenmeyer flask and add 200 ml TAE buffer. Heat in a microwave or on a hot plate to dissolve the agarose. Allow to cool for a few minutes and then add GelRed® Nucleic Acid Gel Stain, 10,000X according to the manufacturer's directions. Once the gel solution is prepared and still a liquid, pour it into the gel tray of the electrophoresis

apparatus. Be sure to place a comb in the gel to create wells for loading samples into the gel. Allow the gel to set and place the tray in the electrophoresis chamber. Add running buffer to the chamber sufficient to cover the gel. The running buffer should be the same as the buffer used to dissolve the agarose.

b) Mix your DNA samples from the PCR run with a gel loading solution in a vol/vol ratio of 9 DNA sample/1 gel loading solution.

c) Load DNA samples from the PCR into the wells in the gel using a micropipette. Normally, 10 µl of the sample should be sufficient. Be sure to include a DNA ladder (a DNA ladder contains DNA fragments of known size and is used to compare with samples to determine the sizes of DNA fragments).

d) Turn on the power and run for about 45 minutes or until the amaranth band is near the end of the gel.

e) Remove and photograph your gel.

18.6 Study Questions

1 Which of your samples, A or B, do you believe contains the Roundup Ready® gene? Recall that one is from conventionally grown (non-organic) corn and the other from organically grown corn. If neither tested positive for Roundup Ready®, explain why this might be given that, presumably, one was Roundup Ready® and the other was organic. If both tested positive, explain how this might have happened.

2 How many bands were visible in each lane of your electrophoresis gel? Was one of them 108 bp in length? If you had more than one band, what might be the identity of the other bands?

3 Do your results from the lateral flow strip and PCR/electrophoresis agree? If not, what might be the explanation?

4 Do you agree with pro-labeling folks that all packaged foods that contain an ingredient from a genetically engineered crop should be labeled "Contains a GMO ingredient?" Why or why not?

18.7 References

1 USDA AMS. Can GMOs be used in organic products? | agricultural marketing service [Internet]. [cited 20 February 2020]. https://www.ams.usda.gov/publications/content/can-gmos-be-used-organic-products

2 USDA ERS. Recent trends in GE adoption [Internet]. [cited 20 February 2020]. https://www.ers.usda.gov/data-products/adoption-of-genetically-engineered-crops-in-the-us/recent-trends-in-ge-adoption.aspx

3 Duke, S.O. and Powles, S.B. (2008). Glyphosate: a once-in-a-century herbicide. *Pest Management Science* 64 (4): 319–325.

4 Pollegioni, L., Schonbrunn, E., and Siehl, D. (2011). Molecular basis of glyphosate resistance – different approaches through protein engineering. *The FEBS Journal* 278 (16): 2753–2766.

5 Swope, N.K., Fryfogle, P.J., and Sivy, T.L. (2015). Detection of the cp4 epsps Gene in Maize Line NK603 and comparison of related protein structures: an advanced undergraduate experiment. *Journal of Chemical Education* 92 (7): 1229–1232.

6 Grothaus, G.D., Bandla, M., Currier, T. et al. (2006). Immunoassay as an analytical tool in agricultural biotechnology. *Journal of AOAC International* 89 (4): 913–928.

7 Namuth, D. (2003). \ [Internet]. Plant & Soil Sciences eLibrary. https://digitalcommons.unl.edu/passel/98

8 DNeasy plant pro kit – QIAGEN [Internet]. [cited 20 February 2020]. https://www.qiagen.com/us/products/discovery-and-translational-research/dna-rna-purification/dna-purification/genomic-dna/dneasy-plant-pro-kit/#orderinginformation

9 Lee, P.Y., Costumbrado, J., Hsu, C.-Y., and Kim, Y.H. (2012). Agarose Gel electrophoresis for the separation of DNA fragments. *Journal of Visualized Experiments* 62: e3923.

10 Romer Labs. AgraStrip® test kits [Internet]. [cited 20 February 2020]. https://www.romerlabs.com/en/products/test-kits/gmo/

11 Datukishvili, N., Kutateladze, T., Gabriadze, I. et al. (2015). New multiplex PCR methods for rapid screening of genetically modified organisms in foods. *Frontiers in Microbiology* [Internet]. [cited 20 February 2020], 6. https://www.frontiersin.org/articles/10.3389/fmicb.2015.00757/full

18.8 Suggested Reading

Chang, Y., Peng, Y., Li, P., and Zhuang, Y. (2017). Practices and exploration on competition of molecular biological detection technology among students in food quality and safety major. *Biochemistry and Molecular Biology Education* 45 (4): 343–350.

DNA Learning Center. Detecting genetically modified foods by PCR [Internet]. [cited 20 February 2020]. http://bioinformatics.dnalc.org/gmo/animation/pdf/Detecting%20GM%20Foods%20by%20PCR.pdf

EnviroLogix. GMO Testing Kits [Internet]. [cited 20 February 2020]. https://www.envirologix.com/gmo-testing/crops-tested/

Roseboro, K. Testing for GMOs [Internet]. The Organic & Non-GMO Report. [cited 20 Februry 2020]. https://non-gmoreport.com/articles/testing-for-gmos/

Ross, J. PCR basics [Internet]. [cited 20 February 2020]. https://www.youtube.com/watch?v=GZBsMxzYacc

19

Food Emulsions and Surfactants

19.1 Learning Outcomes

After completing this exercise, students will be able to:

1) Explain the following terms: emulsion, surfactant, hydrophile–lipophile balance (HLB), amphiphilic.
2) Identify and synthesize common food emulsions.
3) Differentiate between an oil-in-water and a water-in-oil emulsion.
4) Apply the Bancroft Rule when choosing surfactants for food applications.
5) Assess the surfactant solubilization capacity of different dispersion systems.

19.2 Introduction

19.2.1 Emulsions

An emulsion, in its simplest definition, is a mixture consisting of droplets of one immiscible liquid dispersed in a continuous phase of another. In other words, it is a dispersion system in which both the dispersed phase and the continuous phase are liquids. Emulsions are common in foods with oil and water as main components. For example, whole milk can be considered an oil-in-water emulsion in which droplets of milkfat (known as milkfat globules and each surrounded by a thin layer of milkfat globule membrane) are dispersed in the continuous phase of water. Table 19.1 lists some common food emulsions.

Since the two liquids in an emulsion are immiscible, phase separation would tend to occur spontaneously unless a surfactant (also called an emulsifier or emulsifying agent) is present to reduce the interfacial tension. As illustrated in the concise experiment described in Section 5.5.3, a temporary emulsion might be created after a 50:50 mixture of water and vegetable oil is shaken vigorously for 30 seconds, but without a surfactant, phase separation would commence immediately when shaking stops.

19.2.2 Surfactants

Surfactants are amphiphilic molecules that contain both a nonpolar region and a polar region on the same molecule. They are called "surface-active" molecules because of their tendency to

Food Chemistry: A Laboratory Manual, Second Edition. Dennis D. Miller and C. K. Yeung.
© 2022 John Wiley & Sons, Inc. Published 2022 by John Wiley & Sons, Inc.
Companion website: www.wiley.com/go/Miller/foodchemistry2

Table 19.1 Common food emulsions and their typical oil and water contents [1].

Food	Oil content (%)	Water content (%)	Emulsion type
Milk, whole	3.3	88.1	Oil-in-water
Eggnog	4.2	82.5	Oil-in-water
Cream, heavy whipping	36.1	57.7	Oil-in-water
Ranch salad dressing	44.5	45.7	Oil-in-water
Mayonnaise	74.9	21.7	Oil-in-water
Margarine	80.2	17.1	Water-in-oil
Butter	81.1	16.7	Water-in-oil

Table 19.2 Applications of surfactants based on HLB values [3].

HLB value range	Application
3.5–6	w/o (water-in-oil) emulsifier
7–9	Wetting agent
8–18	o/w (oil-in-water) emulsifier
13–15	Detergent
15–18	Solubilizer

concentrate at the interface between two immiscible fluids. For example, in an oil and water mixture, the nonpolar region of the molecule will be attracted to the oil and the polar region to the water. Nonpolar regions are mostly lipophilic hydrocarbon chains or rings, whereas polar regions are hydrophilic functional groups that could be ionic (anionic, cationic, zwitterionic) or nonionic (e.g. a hydroxyl group). The hydrophile-lipophile balance (HLB), which takes into account the size and strength of the hydrophilic groups relative to the lipophilic groups within the molecule [2, 3], is often used for the characterization of surfactants. HLB values typically range from 1 to 20 with a high value indicating a surfactant that is more hydrophilic, and a low value more lipophilic.

Wilder Dwight Bancroft, a chemistry professor at Cornell University, proposed, in a series of papers published between 1912 and 1915, that "hydrophile be used for colloidal solutions in water and hydrophobe for colloidal solutions in non-aqueous solutions" [4–9]. This has been frequently cited as the *Bancroft Rule* when selecting surfactants for different applications. Surfactants with high HLB values of 8–18 (i.e. more hydrophilic) are therefore suitable for oil-in-water emulsions where the continuous phase is water, while surfactants with low HLB values of 3–6 (i.e. more lipophilic) are suitable for water-in-oil emulsions. Table 19.2 provides a broad guideline for the types of surfactant applications based on HLB values [3].

19.2.3 Surfactants in Food Systems

Many food emulsion systems are complex structures with droplets and colloidal particles dispersed in matrices containing components with different physical-chemical properties. Food product developers often include low-molecular-weight (LMW) surfactants in formulations in order to create final products with desirable textural and rheological properties. LMW surfactants (e.g. lecithin), in addition to their surface activity, can interact with food molecules and polymers through intermolecular forces and assemble into three-dimensional gel networks, often referred as liquid crystals or mesomorphic phases – phases intermediate between liquid and crystalline solid [10]. LMW surfactants are widely used in formulating semi-solid foods such as margarine and other fat spreads that require specific structural characteristics. Figure 19.1 shows the structures of some common LMW surfactants. Depending on the surfactant concentration, the water-to-oil ratio, and the presence of stearic acid as a structural component, Gaudino et al. [11] were able to create edible gels with canola oil and water in a network of reverse micellar bundle fibers formed by soy lecithin. Admittedly, the resulting water-in-oil "emulsion" goes beyond the simple definition aforementioned, the end-product, nevertheless, is a homogeneous blend of canola oil and water with a spreadable gel structure aided by lecithin and stearic acid.

Another important use of LMW surfactants in food systems is the dispersion of lipophilic ingredients in aqueous phase. High HLB surfactants such as polysorbates at sufficient concentrations (i.e. above the so-called critical micellar concentration) can aggregate into spherical vesicles

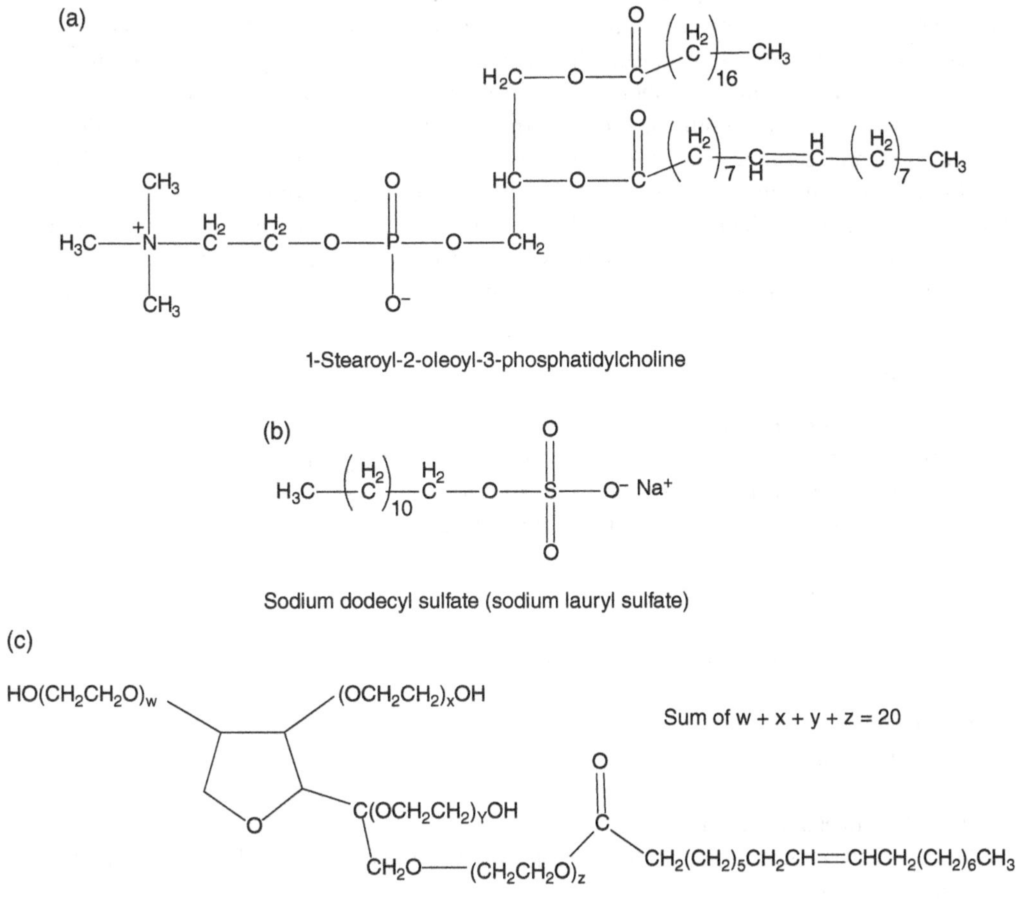

Figure 19.1 Structures of common LMW surfactants. (a) phosphatidylcholine (one of the main phospholipids in soy lecithin); (b) sodium lauryl sulfate; (c) polyoxyethylenesorbitan monooleate, often known as polysorbate 80 or by its trade name TWEEN® 80.

called micelles, which are capable of encapsulating lipophilic ingredients in the center, thereby forming a thermodynamically stable dispersion in water [12]. Surfactant micelles as a delivery system can improve the dispersibility and increase the bioavailability of bioactive compounds [13], making it ideal for new product developments, as well as novel food processing applications. For example, plant-based essential oils with antimicrobial properties have been encapsulated by high HLB surfactant micelles and successfully incorporated into the washing steps for spinach and tomato to reduce pathogenic bacterial loads [14–17].

It should be noted that surfactants are food additives, and therefore their usage is regulated in the United States by the Food and Drug Administration (FDA) [18]. Readers are strongly encouraged to check out the approved surfactant concentrations and uses when developing new product formulations.

19.3 Part I – Butter Churning (Phase Inversion)

Phospholipids and proteins represent about 90% of the dry weight of the milkfat globule membrane, which not only serves as the emulsifying agent for the dispersion of milkfat globules in the aqueous phase of milk but also protects the globules from the attacks by the enzyme lipase [19]. During butter churning, milkfat globules in cream are physically agitated to disrupt the membrane, leading to the coalescence of milkfat and the formation of butter grains, as well as separation from the remaining milk serum called buttermilk. The butter grains are collected and the post-churning working stage further removes remaining buttermilk until a desired consistency is reached. Butter should contain not less than 80% milkfat by weight, with water droplets finely dispersed in the continuous phase of milkfat. Hence, butter churning may be appreciated as a phase-inversion process, as an oil-in-water emulsion (cream) is converted into a water-in-oil emulsion (butter).

19.3.1 Materials and Methods

19.3.1.1 Materials for Buttermaking
1) Glass canning jar with seals and rings, 1-quart, clean (2 per group)
2) Heavy whipping cream at room temperature (125 ml)
3) Heavy whipping cream at refrigeration temperature (125 ml)
4) Teaspoons for kneading butter
5) Bowls for kneading butter
6) Stopwatch
7) Crackers and salt for tasting the butter (optional)
8) Paper plates
9) Plastic knives
10) Conductivity meter

19.3.1.2 Buttermaking Procedure
1) Make sure the jar and its lid are clean. Wash with warm soapy water and rinse thoroughly with cold tap water if needed.
2) Transfer 125 ml of the heavy cream at room temperature into the jar.
3) Close the lid tight and start shaking the jar. **Start the stopwatch.**
4) The cream inside the jar will initially be converted into whipped cream as milkfat starts to clump together and air is being incorporated. Continue to shake the jar even if it feels like the whipped cream is not moving inside the jar.

5) As the agitation continues, the coalescence of milkfat in the cream will eventually lead to the formation of the yellowish butter grains and separation from the remaining milk serum (buttermilk). **Stop the stopwatch when the buttermilk and butter grains separate.** Record the time elapsed.
6) Decant the buttermilk. Replace the lid and shake the jar a few more times to further separate the remaining buttermilk still trapped in the butter grains.
7) Transfer the butter grains to a bowl and use a spoon to gently knead the butter. More buttermilk could be pressed out and should be drained. Repeat this process until a consistent butter is achieved. If you want to taste the butter, transfer a small portion of butter into a paper plate.
8) Clean the jar and the lid with warm soapy water. Rinse thoroughly with cold tap water.
9) Repeat Steps 2–8 with the heavy cream at refrigeration temperature.
10) Store the butter at refrigeration temperature.

19.3.2 Study Questions

1) Weigh the butter and calculate the yield. In other words, how much milkfat from the cream ends up in the butter, assuming the butter contains 80% fat? Did the cream temperature affect the yield? Explain your answer.
2) Did the cream temperature affect the churning time? Explain your answer.
3) What cream temperature treatment is the best for butter churning?
4) Measure the conductivity of cream and butter. Are you able to conclude what type of emulsions they are (oil-in-water or water-in-oil)?

19.4 Part II – Margarine Manufacture (Use of Surfactant for Semi-solid Foods)

Margarine was defined by the US Congress in 1950 (in an Act to repeal the taxes on margarine) to include "all substances, mixtures, and compounds, which have a consistence similar to that of butter and which contain any edible oils or fats other than milk fat if made in imitation or semblance of butter." FDA further set the compositional standard that "margarine (or oleomargarine) is the food in plastic form or liquid emulsion, containing not less than 80% fat" [21CFR166.110]. Although fats from animal origin are allowed, most margarines available in the marketplace are made from vegetable oils. Specifically, partially hydrogenated oils (PHOs), a potential source of harmful *trans* fats, were common ingredients in commercial margarines in the last century. Owing to the increased health concerns of dietary *trans* fats, FDA officially banned the use of PHOs in all food products in the United States starting in June 2018 [20]. The use of surfactants such as lecithin for the formation of semi-solid edible high-fat gels would allow manufacturers to remove PHOs from their formulations, and yet retain the structural characteristics of margarine-type products.

19.4.1 Materials and Methods (Adapted from [11])

19.4.1.1 Materials for Margarine Manufacture
1) Canola oil
2) Water
3) Soy lecithin powder
4) Stearic acid flakes
5) Hotplate, large beaker and boiling chips for setting a boiling water bath

6) Water bath at 55 °C
7) Falcon tubes (or any equivalent plastic tubes)
8) Paper plates
9) Plastic knives
10) Microscope (1000x)

19.4.1.2 Manufacture Procedure

1) Weigh 4 g of lecithin and 4 g of stearic acid.
2) Transfer the lecithin and stearic acid into 32 ml of canola oil in a Falcon tube.
3) Heat the loosely-capped tube in a boiling water bath for 30 minutes to completely dissolve all the ingredients.
4) Move the tube to a water bath at 55 °C.
5) When the oil mixture reaches 55 °C, add 8 ml of water that has also been preheated to 55 °C.
6) Tighten the cap. Shake the tube vigorously for at least one minute.
7) Cool down the tube in a refrigerator. The resulting mixture is margarine.
8) Compare the spreadability of margarine and butter samples (adapted from [21]): Spread some of the sample on fresh bread or a paper plate with a table knife. Spreadability is to be described by one of the following five terms: (i) much too soft, (ii) soft but acceptable, (iii) desirable, (iv) hard but acceptable, and (v) much too hard.
9) Alternatively, measure spreadability with a texture analyzer equipped with the TTC Spreadability Fixture™ (Texture Technologies Corp., Hamilton, MA) by following the manufacturer's procedures. Briefly, the set-up involves an acrylic 90° cone as the penetration probe and the corresponding cone cup sample holders. Samples are first deposited into sample holders and pre-cooled at refrigeration temperature. The maximum peak value in the resulting force-time curve is recorded as spreadability.

19.4.2 Study Questions

1) How would you describe the spreadability of the margarine at refrigeration temperature, as compared to butter? Assuming both butter and margarine samples contain 80% fat, what would your results suggest about the fatty acid composition that lead to the observed difference in spreadability?
2) What other methods could be used by margarine manufacturers to by-pass the use of PHOs in fat spread products?
3) Observe the structure of margarine under a microscope. Would you say the margarine is a water-in-oil emulsion? How do you know?

19.5 Part III – Dispersion of Eugenol in Water (Surfactant Solubilization Capacity)

The solubilization capacity of a surfactant may be estimated by determining the maximum additive concentration (MAC). MAC is the maximum amount of a lipophilic compound that can be solubilized in a micellar solution at a given surfactant concentration [14, 15]. The solubilization capacity is heavily dependent on the surfactant type and increases with surfactant concentration [22]. Environmental factors such as pH [14], temperature [14, 23], and the presence of salt [24] all play a critical role in micellar structure, stability, and properties, and thus could lead to significant variations in the MAC of a surfactant. See the review by Armstrong [25] for a more thorough discussion.

Eugenol, a plant essential oil extracted from clove oil, has been shown to have potent antibacterial and antifungal properties [26, 27]. In this experiment, we will utilize a visible light spectrophotometer to study the solubilization capacity of several high-HLB surfactants when dispersing eugenol in water. At any given surfactant concentration, the MAC will be determined by monitoring changes in optical density (OD) at $\lambda = 632\,nm$ of the surfactant solution with increasing eugenol concentrations. The eugenol concentration at which the absorbance of the surfactant solution starts to deviate from zero, suggesting that no more eugenol molecules could be encapsulated by the surfactant micelles, is noted as the MAC [14, 15].

19.5.1 Materials and Methods (Adapted from [14–16])

19.5.1.1 Materials for Dispersion Experiment
1) Eugenol
2) DI water
3) Polyoxyethylenesorbitan monolaurate or TWEEN® 20; HLB = 16.7
4) Polyoxyethylenesorbitan monooleate or TWEEN® 80; HLB = 15
5) Sodium lauryl sulfate (also called sodium dodecyl sulfate or SDS); HLB = 40 (**Note**: Surfactant HLB values typically range from 1 to 20. Sodium lauryl sulfate is atypical with a value of 40, suggesting that it is *very* hydrophilic.)
6) Beakers for preparing the surfactant stock solutions
7) Falcon tubes (or any equivalent plastic tubes)
8) Vortex mixer
9) Cuvettes
10) Spectrophotometer for OD_{632}

19.5.1.2 Experimental Procedure
1) Prepare a 5% w/v surfactant stock solution by completely dissolving 10 g of surfactant in 190 ml of DI water. (Note: we use 5% as it is said to be a typical surfactant concentration for emulsion applications [14]. Students working in groups are encouraged to try different surfactant concentrations to examine the correlation between MAC and surfactant concentration.)
2) Add eugenol to the surfactant stock solution in tubes at concentrations shown in the following tables.

For TWEEN® 20 and TWEEN® 80 surfactant solutions:

	Amount of eugenol	Amount of surfactant solution	Total amount of the mixture
Eugenol concentration (v/v)	(ml)	(ml)	(ml)
0%	0.00	10.00	10.00
0.5%	0.05 (= 50 µl)	9.95	10.00
0.6%	0.06 (= 60 µl)	9.94	10.00
0.7%	0.07 (= 70 µl)	9.93	10.00

Eugenol concentration (v/v)	Amount of eugenol (ml)	Amount of surfactant solution (ml)	Total amount of the mixture (ml)
0.8%	0.08 (= 80 μl)	9.92	10.00
0.9%	0.09 (= 90 μl)	9.91	10.00
1.0%	0.10 (= 100 μl)	9.90	10.00
1.1%	0.11 (= 110 μl)	9.89	10.00
1.2%	0.12 (= 120 μl)	9.88	10.00
1.3%	0.13 (= 130 μl)	9.87	10.00

For SDS surfactant solution:

Eugenol concentration (v/v)	Amount of eugenol (ml)	Amount of surfactant solution (ml)	Total amount of the mixture (ml)
0%	0.00	10.00	10.00
0.5%	0.05 (= 50 μl)	9.95	10.00
1.0%	0.10 (= 100 μl)	9.90	10.00
1.5%	0.15 (= 150 μl)	9.85	10.00
2.0%	0.20 (= 200 μl)	9.80	10.00
2.5%	0.25 (= 250 μl)	9.75	10.00
3.0%	0.30 (= 300 μl)	9.70	10.00
3.5%	0.35 (= 350 μl)	9.65	10.00
4.0%	0.40 (= 400 μl)	9.60	10.00

3) Mix thoroughly the content of the tubes with the help of a vortex mixer at high setting. You should be able to see a series of samples ranging from clear to turbid as the concentration of eugenol increases.
4) Set the wavelength of the spectrophotometer at 632 nm and first adjust the absorbance to zero using water. Remember to allow time for the spectrophotometer to warm up before making measurements.
5) Use the 0% eugenol sample as the reagent blank. Measure the absorbance of the mixtures at 632 nm.
6) The eugenol concentration at which the absorbance starts to deviate from zero (absorbance greater than 0.05) is the MAC.
7) Repeat Steps 1–6 for another surfactant.

19.5.2 Study Questions

1) Express the solubilization capacity of the surfactant on a per g basis, i.e. ml of solubilized eugenol/g of surfactant. Which surfactant has the highest solubilization capacity?

2) Is the HLB value a good predictor of the solubilization capacity of a surfactant? Discuss other factors that might affect the solubilization capacity.

19.6 Part IV – Mayonnaise Stability

The three basic ingredients for mayonnaise are vegetable oil, vinegar, and egg yolk (with quite a few additives and flavorings being optional). Mayonnaise must contain not less than 65% fat [21CFR169.140], typically 70–75%, but it is an oil-in-water emulsion nonetheless. In other words, mayonnaise is a bit unusual in the sense that the dispersed phase (oil) has a higher concentration than the continuous phase (water). Any emulsifying agents present in the basic ingredients would have to adequately stabilize a large amount of oil in the aqueous phase.

Egg yolk can be separated into two main fractions upon centrifugation: the yellowish liquid plasma and the insoluble granules. Lipoprotein particles, made up of proteins, cholesterol, triglycerides, and phospholipids, are present in both fractions. The plasma fraction on a dry weight basis contains about 85% low-density lipoprotein, while the granules contain about 70% high-density lipoprotein [28]. (**Note**: Densities of lipoproteins vary depending on the relative concentrations of protein and lipids. Densities decrease as lipid concentrations increase and protein concentrations decrease.) Overall, egg yolk contains about 8.2% phospholipids [29] which are the key ingredients that maintain the emulsion stability of mayonnaise.

Salad dressing made with vegetable oil and vinegar is similar to mayonnaise in that egg yolk also serves as the key emulsifying agent, but the fat content is lower. Salad dressing must contain not less than 30% fat [21CFR169.150], typically 40–45%, and includes a starchy paste as a required ingredient. Food gums are optional and often are added by food manufacturers to improve the textural properties of the final product. See Section 5.5.3 for a demonstration of the emulsifying properties of food gums. It is worth noting that commercially available dressings commonly known as salad dressings cannot legally be called "salad dressing" if they do not meet the standard of identity mentioned above [21CFR169.150]. These dressings often do not contain eggs. They are emulsified with gums, often xanthan gum, rather than a phospholipid or other emulsifier. They have names like "Balsamic Vinaigrette Dressing." See the Kraft*Heinz* website for examples of these types of dressings: https://www.myfoodandfamily.com/brands/kraft-dressing/products/25001/products (accessed 03 March 2020).

19.6.1 Materials and Methods

19.6.1.1 Materials for Mayonnaise Experiment
1) Vegetable oil
2) Vinegar (5% acetic acid solution)
3) Liquid egg yolk
4) Water
5) Blender
6) Graduated cylinders, 100 ml

19.6.1.2 Experimental Procedure
Prepare mayonnaise samples at five different egg yolk concentrations following the basic recipes shown in the table below.

Mayonnaise ingredients (ml)	Egg yolk level				
	10%	7.5%	5%	2.5	0%
1) Vegetable oil	75	75	75	75	75
2) Vinegar (5% acetic acid)	15	15	15	15	15
3) Liquid egg yolk	10	7.5	5	2.5	0
4) Water	0	2.5	5	7.5	10

1) Pour the egg yolk, vinegar, and water into a blender. Blend at medium speed for 30 seconds.
2) Add a small portion of the oil to the mixture. Blend at medium speed until the oil is completely mixed in.
3) Repeat Step 2 until all oil is added. Blend at medium speed for an additional 30 seconds.
4) Transfer the mayonnaise into a graduated cylinder. Observe for phase separation after 15 and 60 minutes, and after one, three, and seven days.

19.6.2 Study Questions

1 Vegan mayonnaise contains no egg ingredients. What other natural food ingredients would you recommend to replace egg yolk in vegan mayonnaise? Read the ingredient list of commercial vegan mayonnaise, what ingredients are responsible for the stability of the emulsion?

2 A common optional ingredient in the manufacture of mayonnaise is mustard. Besides contributing to the flavor, what is the purpose of adding mustard? Explain your answer.

3 In addition to phospholipids, egg yolk also contains cholesterol, which has a reported HLB value of 2.7 [30], i.e. it is very lipophilic and thus effective for stabilizing *water-in-oil* emulsions. And yet, mayonnaise is an *oil-in-water* emulsion. Explain. (**Hint**: what is the ratio of phospholipids to cholesterol in egg yolk? And what happens if the ratio is too low?)

19.7 References

1 FoodData Central [Internet]. [cited 29 January 2020]. https://fdc.nal.usda.gov/index.html
2 Griffin, W.C. (1949). Classification of surface-active agents by "HLB". *Journal of the Society of Cosmetic Chemists* 1: 311–326.
3 Davies, J.T. (1957). A quantitative kinetic theory of emulsion type. I. Physical chemistry of the emulsifying agent. *Proceedings of 2nd International Congress Surface Activity*, 426–438.
4 Bancroft, W.D. (1912). The theory of emulsification, I. *The Journal of Physical Chemistry* 16 (3): 177–233.
5 Bancroft, W.D. (1912). The theory of emulsification, II. *The Journal of Physical Chemistry* 16 (5): 345–372.
6 Bancroft, W.D. (1912). The theory of emulsification, III. *The Journal of Physical Chemistry* 16 (6): 475–512.
7 Bancroft, W.D. (1912). The theory of emulsification, IV. *The Journal of Physical Chemistry* 16 (9): 739–758.

8 Bancroft, W.D. (1913). The theory of emulsification, V. *The Journal of Physical Chemistry* 17 (6): 501–519.

9 Bancroft, W.D. (1915). The theory of emulsification, VI. *The Journal of Physical Chemistry* 19 (4): 275–309.

10 Heertje, I., Roijers, E.C., and Hendrickx, H.A.C.M. (1998). Liquid crystalline phases in the structuring of food products. *LWT – Food Science and Technology* 31 (4): 387–396.

11 Gaudino, N., Ghazani, S.M., Clark, S. et al. (2019). Development of lecithin and stearic acid based oleogels and oleogel emulsions for edible semisolid applications. *Food Research International* 116: 79–89.

12 Kralova, I. and Sjöblom, J. (2009). Surfactants used in food industry: a review. *Journal of Dispersion Science and Technology* 30 (9): 1363–1383.

13 Flanagan, J. and Singh, H. (2006). Microemulsions: a potential delivery system for bioactives in food. *Critical Reviews in Food Science and Nutrition* 46 (3): 221–237.

14 Gaysinsky, S., Davidson, P.M., Bruce, B.D., and Weiss, J. (2005). Stability and antimicrobial efficiency of eugenol encapsulated in surfactant micelles as affected by temperature and pH. *Journal of Food Protection* 68 (7): 1359–1366.

15 Gaysinsky, S., Davidson, P.M., Bruce, B.D., and Weiss, J. (2005). Growth inhibition of Escherichia coli O157: H7 and Listeria monocytogenes by carvacrol and eugenol encapsulated in surfactant micelles. *Journal of Food Protection* 68 (12): 2559–2566.

16 Ruengvisesh, S., Loquercio, A., Castell-Perez, E., and Taylor, T.M. (2015). Inhibition of bacterial pathogens in medium and on spinach leaf surfaces using plant-derived antimicrobials loaded in surfactant micelles. *Journal of Food Science* 80 (11): M2522–M2529.

17 Ruengvisesh, S., Oh, J.K., Kerth, C.R. et al. (2019). Inhibition of bacterial human pathogens on tomato skin surfaces using eugenol-loaded surfactant micelles during refrigerated and abuse storage. *Journal of Food Safety* 39 (2): e12598.

18 e-CFR: TITLE 21—Food and Drugs [Internet]. Electronic Code of Federal Regulations. Sect. Part 172—Food Additives Permitted for Direct Addition to Food for Human Consumption. https://www.ecfr.gov/cgi-bin/text-idx?SID=0b8004630f024a1e7e349c48ddc9e30b&mc=true&tpl=/ecfrbrowse/Title21/21cfr172_main_02.tpl

19 Elías-Argote, X., Laubscher, A., and Jiménez-Flores, R. (2013). Dairy ingredients containing milk fat globule membrane: description, composition, and industrial potential. In: *Advances in Dairy Ingredients* (eds. G.W. Smithers and M.A. Augustin), 71–98. Ames, Iowa: Wiley: Institute of Food Technologists. (IFT Press series).

20 FDA (2015). Final determination regarding partially hydrogenated oils. *Federal Register* 80 (116): 34650–34670.

21 Riel, R.R. (1960). Specifications for the spreadability of butter. *Journal of Dairy Science* 43 (9): 1224–1230.

22 Weiss, J. and McClements, D.J. (2000). Mass transport phenomena in oil-in-water emulsions containing surfactant micelles: solubilization. *Langmuir* 16 (14): 5879–5883.

23 Zana, R. and Weill, C. (1985). Effect of temperature on the aggregation behaviour of nonionic surfactants in aqueous solutions. *Journal de Physique Lettres* 46 (20): 953–960.

24 Thévenot, C., Grassl, B., Bastiat, G., and Binana, W. (2005). Aggregation number and critical micellar concentration of surfactant determined by time-dependent static light scattering (TDSLS) and conductivity. *Colloids and Surfaces A: Physicochemical and Engineering Aspects* 252 (2): 105–111.

25 Armstrong, D.W. (1985). Micelles in separations: practical and theoretical review. *Separation and Purification Methods* 14 (2): 213–304.

26 Bullerman, L.B., Lieu, F.Y., and Seier, S.A. (1977). Inhibition of growth and aflatoxin production by cinnamon and clove oils. Cinnamic aldehyde and eugenol. *Journal of Food Science* 42 (4): 1107–1109.

27 Gill, A.O. and Holley, R.A. (2004). Mechanisms of bactericidal action of cinnamaldehyde against listeria monocytogenes and of eugenol against L. monocytogenes and Lactobacillus sakei. *Applied and Environmental Microbiology* 70 (10): 5750–5755.

28 Anton, M. (2013). Egg yolk: structures, functionalities and processes. *Journal of the Science of Food and Agriculture* 93 (12): 2871–2880.

29 Blesso, C.N. (2015). Egg phospholipids and cardiovascular health. *Nutrients* 7 (4): 2731–2747.

30 Pasquali, R.C. and Bregni, C. (2006). Balance hidrofílico-lipofílico (HLB) del colesterol y sus aplicaciones en emulsiones del tipo aceite en agua. *Acta Farmaceutica Bonaerense* 25 (2): 239–244.

19.8 Suggested Reading

Brady, J.W. (2013). *Introductory Food Chemistry*, 638. Ithaca: Comstock Publishing Associates.

Hasenhuettl, G.L. and Hartel, R.W. (eds.) (2008). *Food Emulsifiers and their Applications*, 2e, 426. New York: Springer.

McClements, D.J. (2016). *Food Emulsions: Principles, Practices, and Techniques*, 3e, 690. Boca Raton: CRC Press, Taylor & Francis Group.

Appendix I

Conversion Factors

Mass

1 kg = 10^3 g = 2.2046 pounds (lb)
1 lb= 16 ounces (oz) = 453.59 g
1 oz = 28.35 g
1 ton = 2000 lb = 907.185 kg
1 metric ton = 1000 kg = 1.10 ton

Volume

1 quart (qt) = 4 cups (c) = 32 fluid oz (fl oz) = 0.946 l
1 gallon (gal) = 4 qt = 3.785 l
1 l = 1.0567 qt = 1000 ml
1 c = 8 fl oz = 237 ml =16 tablespoons
1 tablespoon = 3 teaspoons = 0.5 fl oz = 14.8 ml
1 teaspoon = 4.9 ml

Length

1 m= 1.0936 yards = 39.37 inches (in)
1 km = 0.621 miles (mi)
1 mi = 5280 feet (ft) = 1.609 km
1 inch (in) = 2.54 cm
1 Angstrom (Å) = 10^{-10} m

Temperature

°K = °C + 273.15

$$°C = \frac{5}{9} \times (°F - 32)$$

$$°F = \left(\frac{9}{5} \times °C\right) + 32$$

Energy

1 Calorie (Cal) = 1000 cal = 1 kcal
1 cal = 4.184 Joules (J)
1 kcal = 4.184 kJ

Food Chemistry: A Laboratory Manual, Second Edition. Dennis D. Miller and C. K. Yeung.
© 2022 John Wiley & Sons, Inc. Published 2022 by John Wiley & Sons, Inc.
Companion website: www.wiley.com/go/Miller/foodchemistry2

Prefixes

milli $= 10^{-3}$

micro $= 10^{-6}$

nano $= 10^{-9}$

pico $= 10^{-12}$

kilo $= 10^{3}$

Appendix II

Concentration

Concentration is used to express the relative amount of a substance in a solution, food, or other material. When the mixture is homogeneous, as is the case in solutions, the concentration will be the same in all parts of the mixture. Thus, if the concentration of a sample or aliquot taken from a larger portion is measured, the concentration of the entire portion is known. Also, when the concentration and the total volume or weight of a mixture is known, the total amount of the solute can be determined: amount of solute = (concentration) × (volume or weight of solution).

Definition

Concentration is the amount of a substance present per unit volume or weight of a liquid or solid. Units of concentration include two terms arranged in a ratio. Some common expressions of concentration are listed in Table II.1.

Table II.1 Common expressions of concentration.

Unit of concentration	Definition	Common abbreviations
Molarity	mol solute/l solution	M, mol l^{-1}
Millimolarity	mmol solute/l solution	mM, mmol l^{-1}
Normality	Equivalents solute/l solution	N, Eq l^{-1}, equiv l^{-1}
Weight percent	g solute/100 g solution	wt %, w/w %
Milligram percent	mg solute/100 g solution	mg %
Weight/volume percent	g solute/100 ml solution	w/v %
Parts per million	g solute/g solution × 10^6 or mg solute/l solution	ppm, mg kg^{-1}, µg g^{-1}, mg l^{-1}, µg ml^{-1}
Parts per billion	g solute/g solution × 10^9	ppb

Food Chemistry: A Laboratory Manual, Second Edition. Dennis D. Miller and C. K. Yeung.
© 2022 John Wiley & Sons, Inc. Published 2022 by John Wiley & Sons, Inc.
Companion website: www.wiley.com/go/Miller/foodchemistry2

Suggested Reading

Robinson, J.K., McMurry, J., and Fay, R.C. (2019). *Chemistry*, 8e, 1200. Hoboken, NJ: Pearson Education, Inc.

Segel, I.H. (1976). *Biochemical Calculations: How to Solve Mathematical Problems in General Biochemistry*, 2e, 441. New York: Wiley.

Appendix III

Acids, Bases, Buffers, and pH Measurement

Review of pH and Acid–Base Equilibria

Acids and Bases

Recall the following definitions:

Bronsted-Lowry acid: A substance that can donate a proton (H^+).
Bronsted-Lowry base: A substance that can accept a proton.

Acids and bases react to neutralize each other and form salts:

$$HA \ + \ B \ \rightleftharpoons \ HB^+ \ + \ A^-$$

acid   base   conjugate acid   conjugate bass

When the acid donates a proton, the resulting anion becomes a proton acceptor or a base. When the base accepts a proton, it becomes an acid. When chemical species differ only by the presence or absence of a proton, they are called conjugate acid/base pairs: A^- is the conjugate base of HA; BH^+ is the conjugate acid of B. For example, when acetic acid is dissolved in water, it reacts with water (the base in this case):

$$CH_3COOH + H_2O \rightleftharpoons CH_3COO^- + H_3O^+$$

Since the concentration of water is essentially constant in dilute solutions, it is common to write the above equation leaving out the water:

$$CH_3COOH \rightleftharpoons CH_3COO^- + H^+$$

It is important to keep in mind that free protons in water solution are always bound to water molecules. Since some of the acetic acid molecules give up protons in water to form ions, we say that acetic acid dissociates or ionizes in solution.

Acid/Base Equilibria

To understand acid/base chemistry in more depth, we apply the principles of chemical equilibrium. Recall that many chemical reactions do not go to completion, i.e. they are reversible. When this is the case, there will be a point in time after two reactants are brought together when the

Food Chemistry: A Laboratory Manual, Second Edition. Dennis D. Miller and C. K. Yeung.
© 2022 John Wiley & Sons, Inc. Published 2022 by John Wiley & Sons, Inc.
Companion website: www.wiley.com/go/Miller/foodchemistry2

concentrations of reactants and products reach a constant value. This is known as the state of **chemical equilibrium**:

$$aA + bB \rightleftharpoons cC + dD$$

Where A and B are reactants, C and D are products, and a, b, c, and d are coefficients in the balanced chemical equation.

At equilibrium, the concentrations can be described by the equilibrium constant, K_{eq}:

$$K_{eq} = \frac{[C]^c [D]^d}{[A]^a [B]^b}$$

If we can write a balanced equation for a reaction occurring in a solution and if we know the value of the equilibrium constant for the reaction, it is possible to determine the concentration of a given species if the concentrations of the other species are known.

The strengths of acids and bases are determined by the extent to which they ionize in aqueous solution. Acids and bases that ionize nearly completely are termed "strong." HCl and NaOH are examples of a strong acid and a strong base, respectively. Some acids and bases ionize to only a limited extent, making them examples of "weak" acids or "weak" bases. Acetic acid is an example of a weak acid. Ammonia is an example of a weak base. It reacts with water to form an ammonium ion and a hydroxide ion:

$$NH_3 + H_2O \rightleftharpoons NH_4^+ + OH^-$$

The ionization of weak acids and bases at equilibrium may also be expressed as equilibrium constants, in this case referred to as K_a or K_b. For example, the equilibrium expression for ammonia is:

$$K_b = \frac{[NH_4^+][OH^-]}{[NH_3]} = 1.8 \times 10^{-5}$$

Again, the concentration of water is assumed constant and is not included in the equilibrium equation.

Water itself is a weak acid and ionizes to a limited extent:

$$H_2O \rightleftharpoons H^+ + OH^-$$

Thus, the equilibrium constant for water is:

$$K_{eq} = \frac{[H^+][OH^-]}{[H_2O]}$$

Since the concentration of water is assumed constant in aqueous solution, a new equilibrium constant, K_w, is defined:

$$K_w = [H^+][OH^-] = 1 \times 10^{-14}$$

The concentration of H^+ (often written as $[H^+]$) or OH^- ($[OH^-]$) in solutions of weak acids or bases may be calculated if K_a or K_b and the concentration of the acid or base are known. Once $[H^+]$ or $[OH^-]$ is known, the concentration of the other is easily calculated using the equation for K_w.

The pH Scale

Concentrations of hydrogen ions in foods and other biological systems are usually $0.01\,\mathrm{mol\,l^{-1}}$ or less. Therefore, it is often more convenient to express them on a logarithmic scale using pH notation. In this notation, pH is defined as the negative log of the hydrogen ion concentration: $\mathrm{pH} = -\log[H^+]$. To calculate the pH of a system, the concentration of hydrogen ions must first be determined. As an example, let's calculate the pH of pure water. Since $[H^+][OH^-] = 10^{-14}$, and in pure water $[OH^-] = [H^+]$, then $[H^+] = 10^{-7}$ M. Thus, $\mathrm{pH} = -\log 10^{-7} = 7$.

To calculate the pH of a 10^{-2} M solution of a strong acid, HA, we assume complete dissociation, so $[H^+] = [A^-] = 10^{-2}$ M. Thus, $\mathrm{pH} = -\log[H^+] = -\log 10^{-2} = 2$.

Now let's determine the pH of a 10^{-3} M solution of the weak acid HA ($K_a = 1.6 \times 10^{-6}$). At equilibrium:

$$K_a = \frac{[H^+][A^-]}{[HA]} = 1.6\times10^{-6}$$

Since $[H^+] = [A^-]$, we can substitute H^+ for A^-:

$$K_a = \frac{[H^+][H^+]}{10^{-3}-[H^+]} = 1.6\times10^{-6}$$

Since HA is a weak acid, we can assume that $[H^+]$ is very much smaller than $[HA]$ (why?) and thus we can disregard it when calculating the concentration of HA in solution. Solving for $[H^+]$, we have $[H^+] = 4.0 \times 10^{-5}$. To calculate pH, we simply plug into the equation for pH:

$$\mathrm{pH} = \log\frac{1}{[H^+]} = \log\frac{1}{4.0\times10^{-5}} = \log(2.5\times10^4) = \log 2.5 + \log 10^4 = 4.4$$

Refer to any introductory chemistry book or see Segel [1] for exercises of this kind.

pK

Another important concept to review is pK. K_a or K_b are frequently expressed in terms of pK instead of K. $\mathrm{p}K_a = -\log K_a$ and $\mathrm{p}K_b = -\log K_b$. Thus, an acid with a K_a of 10^{-3} has a $\mathrm{p}K_a$ of 3. What is the $\mathrm{p}K_b$ of a base with $K_b = 1.4 \times 10^{-2}$?

The $\mathrm{p}K_a$ of an acid may be determined by titrating the acid with base and plotting a titration curve. The titration graph of a weak acid is shown in Figure III.1. Notice that in the region between pH 4 and 6, large amounts of base are added with little change in pH. This, by definition, is the buffer region. After pH 6.5, tiny amounts of base produce a very large change in pH. This is termed the equivalence point, where all acid molecules have been converted to the salt form, i.e. the number of equivalents of base added equals the number of equivalents of acid initially present. To determine the pK, simply find the pH at which half the base equivalents have been added, in this case 500 ml. At this point, $[HA] = [A^-]$, and so $\mathrm{pH} = \mathrm{p}K_a$.

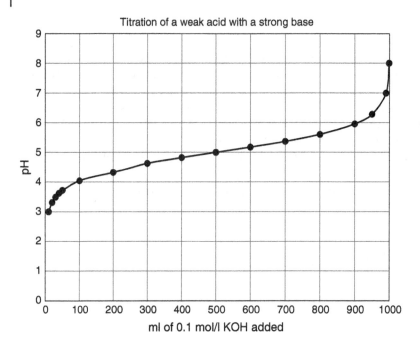

Figure III.1 Titration of a weak monoprotic acid, HA (pK$_a$ = 5.00) with a strong base (KOH). Note that the acid is half titrated, [HA = A$^-$], when 500 ml of the base is added.

Buffers: Functions and Uses

The addition of even small amounts of strong acids or bases to water or to dilute solutions of strong acids or bases causes large changes in pH. The pH of a buffered solution, on the other hand, changes only a little when small amounts of acids or bases are added. Buffers are solutions containing a weak acid and its conjugate base or a weak base and its conjugate acid.

To understand how buffers work, recall that the K$_a$ of an acid is defined as:

$$K_a = \frac{[H^+][A^-]}{[HA]}$$

Rewriting we have:

$$[H^+] = K_a \frac{[HA]}{[A^-]}$$

Taking the negative log of both sides:

$$-\log[H^+] = -\log K_a - \log\frac{[HA]}{[A^-]}$$

Or

$$pH = pK_a + \log\frac{[A^-]}{[HA]}$$

This is the familiar Henderson–Hasselbalch equation. Notice that when the concentrations of the acid HA and its salt, A^-, are equal, $pH = pK_a$.

The Henderson–Hasselbalch equation is useful for a number of reasons. First, it may be used to show that additions of small amounts of acid or base to a buffer result in very small changes in pH compared to what would occur in pure water or in a solution of the salt of a strong acid and a strong base such as NaCl. It also allows an easy way to calculate the amounts of acid and base to add to a solution to obtain a desired pH (in other words, how to make up a buffer.) Finally, it may be used to calculate pK from experimental data.

To see how buffers stabilize pH, let us compare the effects of adding 100 ml of 0.01 M HCl to 1.00 l of water at pH 7 and to 1.00 l of a 1.0 M pH 7 buffer (HA/A^-). Assume the pK_a of HA equals 7.0. (Note that 100 ml of 0.01 M HCl contains 0.001 mol of HCl.) In the water–acid mixture, the hydrogen ion concentration would be: $0.001\,mol/1.1\,l = 0.000909\,mol\,l^{-1}$. Therefore, pH may be calculated as follows:

$$pH = -\log 9.09 \times 10^{-4} = -0.96 + 4 = 3.04$$

When acid is added to the buffer, most of the protons from the HCl would combine with A^-:

$$HCl + A^- \rightleftharpoons HA + Cl^-$$

Thus, to calculate the concentration of H^+ in the buffer after the HCl addition, we must account for the protons that combined with A^-. Remember that at pH 7, $[HA] = [A^-] = 0.5$ M since the pK_a of HA equals 7.0. (**Note:** The concentration of a buffer equals the concentration of the acid plus the concentration of its conjugate base.) Adding 0.001 mol HCl to the system gives $[HA] = 0.501\,mol/1.1\,l$ and $[A^-] = 0.499\,mol/1.1\,l$. Using the Henderson/Hasselbalch equation:

$$pH = 7.00 + \log \frac{0.499/1.1}{0.501/1.1} = 7.00 - 0.002 = 6.998$$

Clearly, the buffer serves to inhibit the large change in pH that occurs when acid is added to water.

A third way in which the Henderson–Hasselbalch equation is useful is in making up buffers at a given pH and in determining buffer capacity. Buffer capacity may be defined in two ways: (i) the number of moles H^+ or OH^- required to give a certain pH change in 1 l of a buffer or (ii) the pH change that occurs when a given amount of acid or base is added to 1 l of the buffer. The buffer capacity is a function of the concentration of the buffer. The higher the concentration, the higher the buffer capacity.

Problems

1   How much acetic acid ($pK_a = 4.75$, $K_a = 1.6 \times 10^{-5}$) and sodium acetate are required to make 1 l of 0.2 M acetate buffer, pH 4.5?

Let HA = acetic acid, and A^- = acetate. In a 0.2 M buffer, $[HA] + [A^-] = 0.2$ M;

Let $[A^-] = x$; then $[HA] = 0.2-x$;

Plugging into the Henderson–Hasselbalch equation:

$$4.5 = 4.75 + \log\frac{x}{0.2 - x}$$

$$0.25 = \log\frac{0.2 - x}{x}$$

$$1.78 = \frac{0.2 - x}{x}$$

$$x = 0.07$$

Therefore, for a 0.2 M buffer at pH 4.5, add 0.13 moles of acetic acid and 0.07 moles of sodium acetate and dilute to 1 l.

Now you calculate the amounts of acetic acid and sodium acetate necessary for preparing 1.0 l of a 0.4 M acetate buffer, pH 4.5. **Answers:** 0.256 moles of acetic acid and 0.144 moles of sodium acetate.

2   How much 1.0 M HCl is required to change the pH of 1 l of a 0.2 M acetate buffer, pH 4.5, to 3.5? (The answer is the buffering capacity of the buffer.)

To solve the problem, we simply calculate the concentration of HA at pH 3.5 and at pH 4.5 and subtract.

Plugging into the Henderson–Hasselbalch equation:

$$3.50 = 4.75 + \log\frac{x}{0.2 - x}$$

$$\log\frac{0.2 - x}{x} = 1.25$$

$$18.78 x = 0.2$$

$$x = 0.01$$

Thus, at pH 3.5, the concentration of HA = 0.19 M and the concentration of A⁻ = 0.01 M. Earlier, we found that [HA] at pH 4.5 was 0.13 M. Thus, buffering capacity in the acid direction equals 0.19−0.13 = 0.06 mol l⁻¹.

The buffer capacity in the alkaline direction would be calculated as follows:

$$5.5 = 4.75 + \log\frac{x}{0.2 - x}$$

$$x = 0.17$$

Thus, [A⁻] = 0.17 and buffer capacity equals 0.17−0.07 = 0.1 mol l⁻¹.

Now you calculate buffer capacity for the 0.4 M buffer. **Answers**: Acid, 0.123 mol l⁻¹; Base, 0.196 mol l⁻¹.

Choosing a Buffer System

There are many buffer systems available and choosing the best buffer is not always easy. Clearly, the most important criteria in choosing an appropriate buffer is to match the pH desired with the pK of the buffer. The closer the pK is to the desired pH, the more buffering capacity the buffer will have. Generally, the rule of thumb is that the ratio of conjugate base concentration to acid concentration should be between 0.1 and 10.0 for effective buffering i.e. buffers work best at pHs within $\pm$ 1 unit of the buffer pK. A few simple calculations using the Henderson-Hasselbalch equation will show you how much buffering capacity exists within and outside this range.

Acetic, citric and phosphoric acids along with their sodium or potassium salts are widely used buffering agents in food systems. The following short list shows the effective buffering ranges of these salts:

Buffer system	Effective buffering range, pH
Citric acid–sodium citrate	2.1–4.7
Acetic acid–sodium acetate	3.6–5.6
Ortho and pyrophosphate anions	2.0–3.0, 5.5–7.5, and 10–12

When used for experimental purposes, a buffer should function to maintain the pH at a specified value and should not influence experimental variables. Thus, it is important to choose a buffer which does not interact with the other substances being used in the experiment. Some decades ago, an extensive series of studies was done on available buffers by Good and his coworkers [2]. They set forth a list of criteria for choosing a buffer and then went about designing new buffers or evaluating old ones based on these criteria. The so-called Good buffers are now widely used in food chemistry, biochemistry, and biological research. Good's criteria along with brief explanations follow.

1) pK_a between 6 and 8. This criteria is more useful for biochemistry experimentation and less useful for food systems; Good's point was that biochemists would profit from having a number of buffers operating in the range of interest, rather than relying on one buffer of $pK_a = 7$.
2) High solubility in aqueous systems. Clearly important in biology, although not always important in foods.
3) Exclusion by biological membranes.
4) Minimal salt effects. It is important that salts and other ions in the system do not affect the buffer and that the buffer does not act as an interfering ion in the experiment.
5) Little influence of concentration, temperature, or ionic strength of the solution on degree of dissociation. The buffer should be able to maintain a pH level through changes in storage temperature or through addition of salts or other components of a system. Thus, minor changes can be made in product formulations without requiring changes in the buffer. In practice, no buffer is totally unaffected by these influences.
6) Well-defined or nonexistent interactions with mineral cations. Many acids or salts will bind with metal ions, thus changing the dissociation equilibrium of the acid and reducing the functional use, if any, of the metal.

7) Chemical stability. This is of obvious importance. Buffers must not break down with storage or light exposure.
8) Insignificant light absorption in the visible range. This characteristic is important in experiments where spectrophotometric measurements are planned.
9) Easily available in pure form. For food systems, this should be extended to include availability in Food Grade.

Some common buffers along with pKa values are listed in Table III.1.

Table III.1 pKa values for some common biological buffers [3].

Name	Chemical name	pKa
Phosphate		2.12 (pKa$_1$)
Citrate		3.06 (pKa$_1$)
Formate		3.75
Succinate		4.19 (pKa$_1$)
Citrate		4.74 (pKa$_2$)
Acetate		4.75
Citrate		5.40 (pKa$_3$)
Succinate		5.57 (pKa$_2$)
MES	2-(N-morpholino)ethanesulfonic acid	6.15
ADA	N-2-acetaminoiminodiacetic acid	6.60
BIS-TRIS propane	1,3-bis(hydroxymethyl)methylamino propane	6.80
PIPES	piperazine-N,N'-bis(2-ethanesulfonic acid)	6.80
ACES	N-2-acetamido-2-aminoethanesulfonic acid)	6.90
Imidazole		7.00
MOPS	3-(N-morpholino)propanesulfonic acid	7.20
Phosphate		7.21 (pKa$_2$)
TES	N-trishydroxymethylmethyl-2-aminoethanesulfonic acid	7.50
HEPES	N-2-hydroxyethylpiperazine-N-2-ethanesulfonic acid	7.55
HEPPS	N-2hydroxyethylpiperazine-N'-3propane sulfonic acid	8.00
TRICINE	N-tris(hydroxymethyl)methylglycine	8.15
Glycine amide, hydrochloride		8.20
TRIS	Tris(hydroxymethyl)aminomethane	8.30
BICINE	N,N-Bis(2-Hydroxyethyl)glycine	8.35
Glycylglycine		8.40
Borate		9.24
CHES	Cyclohexylaminoethanesulfonic acid	9.50
CAPS	3-(cyclohexylamino)propanesulfonic acid	10.40
Phosphate		12.32 (pKa$_3$)

Preparation of Buffers

To prepare a buffer, you can either start with both the acid and its conjugate base and calculate the amounts of each to add, or you can start with either the acid or the conjugate base, adjust the pH with a strong acid or base, and then make to volume. In either case, the pH should be checked and adjusted before bringing the buffer to the final desired concentration. Buffers will actually vary slightly from theoretical pH. Thus, it is better to use standardized formulae for making up buffers of a desired pH. These may be found in the CRC Handbook of Biochemistry and Molecular Biology [4] and other handbooks. Also, Sigma-Aldrich has a "Buffer Reference Center" on their website [5] that may be helpful. It includes tables for preparing some frequently used buffers. Directions for preparing some common buffers are given in Tables III.2, III.3, and III.4.

Table III.2 Some standard buffers, pH ranges where they are effective buffers, and effects of temperature on pH[a,b].

Buffer no.	Name of buffer	pH range	Temp	ΔpH/°K
1	Glycine/HCl	1.2–3.4	Room	0
2	Na citrate/HCl	1.2–5.0	Room	0
3	Na citrate/NaOH	5.2–6.6	20°C	+0.004
4	Phosphate	5.0–8.0	20°C	0.003
5	Citric acid/phosphate	2.2–7.8	21°C	
6	Acetate	3.5–5.6	25°C	
7	Tris maleate	5.2–8.6	23°C	
8	Tris/HCl	7.2–9.0	23°C	

[a] Adapted from [6].
[b] See [4] for a more extensive list of buffers.

Table III.3 Stock solutions and formulas for mixing buffers.

	Stock solutions for preparing buffers[b]		Mixing formulas for
Buffer no.[a]	A	B	preparing buffers[c]
1	0.1 M glycine + 0.1 M HCl	0.1 M HCl	x ml A + (100−x) ml B
2	0.1 M disodium citrate (21.01 g citric acid monohydrate + 200 ml 1 M NaOH diluted to 1.0 l)	0.1 M HCl	x ml A + (100−x) ml B
3	0.1 M disodium citrate (21.01 g citric acid monohydrate + 200 ml 1 M NaOH made to 1.0 l)	0.1 M NaOH	x ml A + (100−x) ml B
4	1/15 M potassium dihydrogen phosphate	1/15 M disodium phosphate	x ml A + (100−x) ml B
5	0.1 M citric acid	0.2 M disodium phosphate	x ml A + (100−x) ml B

(Continued)

Table III.3 (Continued)

Buffer no.[a]	Stock solutions for preparing buffers[b]		Mixing formulas for preparing buffers[c]
	A	B	
6	0.1 M sodium acetate	0.1 M acetic acid	x ml A + (100−x) ml B
7	0.2 M Tris maleate (24.23 g Tris + 23.21 maleic acid made to 1.0 l)	0.2 M NaOH	25 ml A + x ml B made up to 100 ml
8	0.2 M Tris (24.23 g Tris made to 1.0 l)	0.1 M HCl	25 ml A + x ml B made up to 100 ml

[a] See Table III.2 for the name of the buffer.
[b] Stock solutions should be prepared from CO_2-free distilled water and reagent-grade chemicals.
[c] Values for x are given in Table III.4.

Table III.4 Values for x (see Tables III.2 and III.3) to make up various buffers of a specified pH.

pH	1	2	3	4	5	6	7	8	pH
1.2	11.1	9.0							1.2
1.4	26.4	17.9							1.4
1.6	36.2	23.6							1.6
1.8	43.9	27.6							1.8
2.0	50.7	30.2							2.0
2.2	56.5	32.2			98.8				2.2
2.4	62.3	34.1			94.5				2.4
2.6	68.4	36.0			90.0				2.6
2.8	74.4	37.9			85.1				2.8
3.0	81.0	39.9			80.3				3.0
3.2	86.2	42.1			76.0				3.2
3.4	90.3	44.8			72.0				3.4
3.6		47.8			68.4				3.6
3.8		51.2			65.1	10.9			3.8
4.0		55.1			62.0	16.6			4.0
4.2		60.0			59.1	23.9			4.2
4.4		66.4			56.4	33.5			4.4
4.6		74.9			53.7	44.9			4.6
4.8		85.6			51.2	56.6			4.8
5.0		100.0		99.2	49.0	67.8			5.0
5.2			87.1	98.4	46.9	76.8	3.2		5.2
5.4			78.0	97.3	44.7	84.0	5.0		5.4

Table III.4 (Continued)

pH	1	2	3	4	5	6	7	8	pH
					Buffer no.				
5.6			70.3	95.5	42.4	89.3	7.3		5.6
5.8			64.5	92.8	40.0		9.7		5.8
6.0			60.3	88.9	37.4		12.4		6.0
6.2			57.2	83.0	34.5	ı	15.2		6.2
6.4			54.8	75.4	31.4		17.9		6.4
6.6			53.2	65.3	27.9		20.8		6.6
6.8				53.4	23.5		22.2		6.8
7.0				41.3	19.0		23.7		7.0
7.2				29.6	13.8		25.2	44.7	7.2
7.4				19.7	9.8		26.7	42.0	7.4
7.6				12.8	6.8		28.6	39.3	7.6
7.8				7.4	4.6		31.2	33.7	7.8
8.0				3.7			33.9	27.9	8.0

Activity and Ionic Strength

When working buffer problems with the objective of understanding how buffers work, we generally assume that buffers are ideal solutions. In practice, such an assumption can get you into trouble. This is because some other characteristics of a buffer solution that we haven't considered so far may affect the pH of the buffer and its buffering capacity. These are concentration (adding a very dilute acid or base will greatly increase total volume, changing the extent of dissociation of the buffer), ionic strength, and temperature. Also, if the buffer used is polyprotic, such as phosphate or citrate, then it will have more than one pK and will buffer in more than one region.

In order to understand these factors, we need to review the concepts of activity and ionic strength. As solutions become more concentrated, solutes begin to interact in ways that reduce their "effective concentrations." When this happens, the participation of a solute in an equilibrium reaction will be less than predicted from its actual concentration. Activity is a term for concentration that reflects the *effective* concentration of a solute. It is related to concentration as follows:

$$a = f\left[A\right]$$

where: a = activity of A
   f = the activity coefficient
   $[A]$ = the concentration of A in moles/liter

The activity coefficient is usually less than 1 and it approaches 1 as the concentration of A approaches zero, i.e. in very dilute solutions activity equals concentration. Activity coefficients are related to ionic strength as shown in Figure III.2.

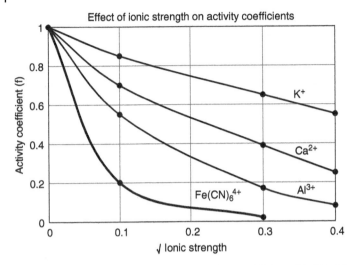

Figure III.2   Effect of ionic strength on activity coefficients. Modified from [7].

Ionic strength is an expression of the concentration of ions in solution.

$$I = \frac{1}{2}\sum c_i z_i^2$$

Where: I = ionic strength
$\sum$ = the "sum of"
c_i = the molar concentration of the ith ion
z_i = the charge on the ith ion (1, 2, etc.)

The Henderson–Hasselbalch equation states that the pH of a buffer is dependent only on the ratio of the concentration of the conjugate base to the conjugate acid. Actually pH depends on the ratio of *activities* of these species. Unfortunately, the effect of ionic strength on the activity of a solute is not the same for all ions. The greater the charge on an ion, the greater will be the effect of increases in ionic strength on its activity. Table III.5 lists the activity coefficients of some ions in solution.

Ions other than the buffer ions will also affect the activities of the buffer components. For example, addition of NaCl to a buffer solution will affect the pH of the solution because of its contribution to the total ionic strength of the solution. In this case, it is common to correct the pK rather than the activities of the buffer ions. Details of this type of correction may be found in Segel [1].

pH Measurement

pH can be measured using either pH indicators or a pH meter. Indicators have been widely used, but their utility in food systems is limited by the fact that foods are usually opaque and colored themselves and changes in indicator color are not readily seen. Also, indicators are not very sensitive and provide only a ballpark indication of pH. Thus, pH values are usually measured using pH meters, which accurately measure the hydrogen ion activity of solutions.

Table III.5 Values for activity coefficients of some ions at different concentrations in aqueous solution [1].

Ion	Ionic concentration (M)		
	0.001 M	0.01 M	0.1 M
H^+	0.975	0.933	0.86
OH^-	0.975	0.925	0.805
Acetate$^-$	0.975	0.925	0.82
H_2PO^-	0.975	0.928	0.744
HPO_4^{2-}	0.903	0.740	0.445
PO_4^{3-}	0.796	0.505	0.16
$H_2Citrate^-$	0.975	0.926	0.81
$HCitrate^{2-}$	0.903	0.741	0.45
$Citrate^{3-}$	0.796	0.51	0.18
HCO_3^-	0.975	0.928	0.82
CO_3^{2-}	0.903	0.742	0.445

Detailed descriptions of how pH meters operate can be found in most books on analytical chemistry. The following is a brief discussion of the operation and care of pH meters.

The pH meter consists of a voltmeter and two electrodes, the reference electrode and the indicator electrode. Frequently, the two electrodes are physically combined into a combination electrode, which looks like a single electrode. The indicator electrode contains a glass membrane, which is permeable to H^+ but not other ions. Thus, hydrogen ions in the solution can diffuse across the membrane unaccompanied by anions. The accumulated hydrogen ions inside the electrode cause a positive charge to be built up with a negative charge on the other side, thus an electric potential is set up across the glass membrane. This potential can be measured by the voltmeter due to the circuit created by the indicator electrode, the reference electrode, and the solution. The reference electrode has a porous plug, which allows contact with the solution and completes the circuit.

If you are interested in the details of pH meter function, you should consult one of the references listed. pH meters are simple to use and give accurate results if the following steps are followed and appropriate precautions are taken.

Making pH Measurements

1) The pH meter must be calibrated before use with a buffer of appropriate pH. Buffer standards may be purchased from chemical supply houses. The pH of the buffer standard should be as close as possible to the pH to be measured.
2) It is important that the temperature control unit be adjusted to match the temperature of the solution being measured.
3) The electrode must be treated with care!! Always leave it in water or some solution to prevent it from drying out and be careful not to scratch or chip the glass bulb at the bottom. Before immersing into a solution, the electrode should be rinsed and blotted dry. The electrode should not touch the bottom or sides of the container when measuring pH.

4) The electrode must be kept clean. This can be a challenge, especially when solutions containing proteins, lipids, and viscous carbohydrate are used. Cleaning procedures for electrodes are provided by the manufacturer and should be consulted. In some cases, soaking in detergent solutions or rinsing with acetone will do the trick. Sluggishness or drift in the pH reading is an indication that the electrode may be dirty.

References

1 Segel, I.H. (1976). *Biochemical Calculations: How to Solve Mathematical Problems in General Biochemistry*, 2e, 441. New York: Wiley.
2 Good, N.E., Winget, G.D., Winter, W. et al. (1966). Hydrogen ion buffers for biological research. *Biochemistry* 5 (2): 467–477.
3 Mohan, C. (2003). *Buffers – A Guide for the Preparation and Use of Buffers in Biological Systems.* Calbiochem: EMD Biosciences Inc.
4 Lundblad, R.L. and Macdonald, F. (eds.) (2018). *Handbook of Biochemistry and Molecular Biology*, 5e, 1001. Boca Raton: CRC Press, Taylor & Francis Group.
5 Buffer Reference Center | Sigma-Aldrich [Internet]. [cited 10 February 2020]. https://www.sigmaaldrich.com/life-science/core-bioreagents/biological-buffers/learning-center/buffer-reference-center.html
6 Lentner, C. (ed.) (1984). *Geigy Scientific Tables. 3: Physical Chemistry, Composition of Blood, Hematology, Somatometric Data*, 8th, rev.enl.e, 359. Basel: CIBA-Geigy AG.
7 Skoog, D.A., Holler, F.J., and Crouch, S.R. (2018). *Principles of Instrumental Analysis*, 7e, 959. Australia: Cengage Learning.

Suggested Reading

Lindsay, R.C. (2017). Food additives. In: *Fennema's Food Chemistry*, 5e (eds. S. Damodaran and K.L. Parkin), 803–864. Boca Raton: CRC Press, Taylor & Francis Group.
Miller, D.D. (2017). Minerals. In: *Fennema's Food Chemistry*, 5e (eds. S. Damodaran and K.L. Parkin), 627–679. Boca Raton: CRC Press, Taylor & Francis Group.
Robinson, J.K., McMurry, J., and Fay, R.C. (2019). *Chemistry*, 8e, 1200. Hoboken, NJ: Pearson Education, Inc.
Skoog, D.A., West, D.M., Holler, F.J., and Crouch, S.R. (2012). *Fundamentals of Analytical Chemistry*, 9e, 1072. Belmont, CA: Cengage Learning.

Appendix IV

Spectrophotometry

Introduction

Note: The following is a brief overview of the basic principles of spectrophotometry. For an excellent and detailed description of UV/visible spectrophotometry, as well as general spectroscopy, see Penner [1, 2].

Electromagnetic radiation may be classified according to wavelength or frequency. Frequency and wavelength are related according to the following formula:

$$\lambda v = c \tag{IV.1}$$

where λ is the wavelength, v is the frequency, and c is the speed of light (3×10^{10} cm s^{-1}). Units for wavelength are usually nanometers (nm), but Angstroms (Å), microns (μm), or centimeters (cm) may also be used.

$1\,\text{Å} = 10^{-10}\,\text{m}$
$1\,\text{nm} = 10^{-9}\,\text{m}$
$1\,\mu\text{m} = 10^{-6}\,\text{m}$
$1\,\text{cm} = 10^{-2}\,\text{m}$

Table IV.1 shows wavelength ranges for the various classifications of electromagnetic radiation.

Chemical species interact with electromagnetic radiation in ways that reduce the intensity of the radiation. The effects of the interaction vary with the wavelength of the radiation and the nature of the chemical species and include transitions in molecular vibrations, molecular rotations, and electron energy levels. Interactions may be quantified by comparing the radiation incident on a sample with the radiation transmitted by the sample. Such comparisons are generally made with instruments called spectrophotometers and require that the sample be dissolved in a solvent that does not absorb appreciably at the wavelength of interest.

Spectrophotometers generally follow the basic setup shown in Figure IV.1. The essential elements and their functions are as follows:

1) *Light source.* Usually a tungsten filament lamp is used to generate wavelengths from 340 to 900 nm (visible) and a hydrogen lamp is used to generate wavelengths from 200 to 360 nm (UV).
2) *Monochromator.* Since it is desirable to have monochromatic light incident on the sample, some method must be used to obtain a narrow range of wavelengths from the polychromatic light generated by the lamps. Prisms or diffraction gratings are used to separate the different wavelengths into narrow ranges.

Food Chemistry: A Laboratory Manual, Second Edition. Dennis D. Miller and C. K. Yeung.
© 2022 John Wiley & Sons, Inc. Published 2022 by John Wiley & Sons, Inc.
Companion website: www.wiley.com/go/Miller/foodchemistry2

Table IV.1 Classification of electromagnetic radiation and effects of radiation on molecules.

Region	X-rays	Ultraviolet	Visible	Infrared	Microwave
Wavelength, nm	0.1–100	100–400	400–800	800–100,000	10^5–10^8
Effect on molecule	Excitation of subvalence electrons	Excitation of valence electrons	Excitation of valence electrons	Increased molecular vibrations	Increased molecular rotations

Source: Modified from [3].

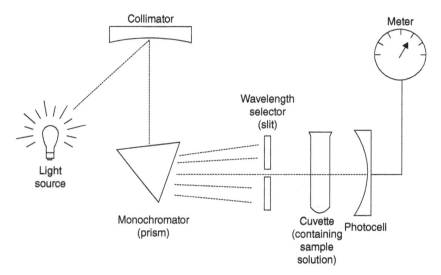

Figure IV.1 Schematic diagram of the basic components of a spectrophotometer. *Source:* Modified from [3].

3) *Slit.* Light leaving the monochromator is not truly monochromatic but consists of a narrow range of wavelengths. The purpose of the slit is to decrease the width of this band to improve the purity of the light that strikes the sample.
4) *Sample containers.* Generally 1-cm-wide rectangular cuvettes are used.
5) *Light-detecting phototube or photocell.* These are devices that contain a photosensitive material that emits electrons when exposed to light. The electrons are emitted in an amount proportional to the light intensity. Thus, the intensity of the transmitted light I can be measured.

The amount of light that interacts with a sample in a spectrophotometer may be expressed as either transmittance or absorbance. Transmittance T is the fraction of light transmitted by a solution:

$$T = I / I_0 \tag{IV.2}$$

where I is the intensity of the radiation transmitted by the sample, and I_0 is the intensity of the radiation incident on the same sample. Absorbance A is an expression of the amount of radiation absorbed by a sample and is equal to the negative log of T:

$$A = -\log T = \log I_0 / I \tag{IV.3}$$

The amount of monochromatic radiation that is absorbed by a solution is proportional to the concentration of the absorbing species in the solution and the distance the radiation travels through the sample. Beer's law is an expression of these relationships:

$$A = \varepsilon bc \tag{IV.4}$$

where ε is the absorption coefficient (also called the molar extinction coefficient), b is the path length of the radiation in the sample, and c is the concentration of the absorbing species. The units most used for each of these are shown below:

A: absorbance has no units
ε: $M^{-1}cm^{-1}$ or $mM^{-1}cm^{-1}$
b: cm (cuvettes with 1 cm path lengths are the most common)
c: M or mM

Beer's law is useful for determining concentrations of solutes in solution. It states that there should be a linear relationship between concentration and absorbance. Therefore, if ε and b are known, it is possible to calculate concentration from an absorbance value measured with a spectrophotometer. Of course, one must be certain that the wavelength setting and the absorbance reading of the spectrophotometer are accurate and that there are no interfering substances present. In practice, it is usually better to run a series of standards of known concentration and determine the concentration of the unknown from a standard curve.

In many cases, the substance to be measured does not absorb appreciably at a suitable wavelength but may react with a second substance to form a colored compound that does absorb. When carrying out this type of an assay, it is important that the second substance be present in excess, so it is not limiting for color formation. A general equation for this type of assay is shown below:

$$\text{Substance to be assayed}\left(\text{substance A}\right) + \text{Color-forming reagent}\left(s\right) \rightarrow$$
$$\text{Substance that absorbs in proportion to concentration of substance A} \tag{IV.5}$$

When using spectrophotometry to determine the concentration of a substance in solution, it is best to construct a standard curve. This is done by preparing a series of standards of varying concentrations and reacting the standards with the color-forming reagents under the exact conditions that will be used when assaying the unknown. In most cases, the color-forming reagents and solvents will absorb a small amount at the wavelength of interest. It is a good idea, therefore, to prepare a reagent blank that contains all of the reagents at the appropriate concentrations but none of the substance to be assayed. The absorbance of the reagent blank can be subtracted from that of the sample by either zeroing the spectrophotometer with the reagent blank or zeroing with the solvent and subtracting the absorbance of the reagent blank from each sample absorbance reading. Absorbance measurements are most accurate in the range of 0.1–1.0 absorbance units. It is wise, therefore, to adjust concentrations (by dilution or adding more sample) so that absorbance readings will be in this range. Concentration adjustments should be made *before* the color-forming reagents are added.

Apparent deviations from Beer's law do occur, and it is important to be aware of situations in which nonlinearity may be a problem. Beer's law holds only for dilute solutions. In concentrated solutions, molecules may interact in ways that change their absorptive properties. If your standard curve includes concentrations both higher and lower than that of your unknown sample, then you will know if you have a problem and can make adjustments accordingly. Another common cause of nonlinearity is insufficient amounts of color-forming reagents. Figure IV.2 shows a standard curve.

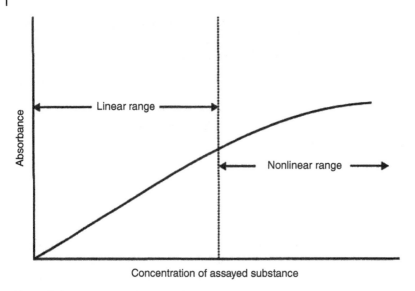

Figure IV.2   A typical standard curve obtained in quantitative spectrophotometry.

Operation of a Spectrophotometer

While key components of most spectrophotometers are similar, operating procedures provided by manufacturers might vary. It is worthwhile to carefully review the instruction manual of the instrument you are using to ensure proper operation and maintenance.

Notes for Operators

1) Allow time for the spectrophotometer to warm up before starting your experiment.
2) When operating the instrument over a period of time, recheck 100% T periodically and reset if necessary.
3) The 100% T adjustment must be reset every time the wavelength is changed.
4) All solutions must be free of bubbles.
5) Cuvettes must be at least half full.
6) Cuvettes must be clean. Use a low-lint antistatic tissue to wipe off fingerprints.
7) Align the mark on the cuvettes with the direction of the monochromatic light beam.
8) Check for any spills or drips after each measurement. The sample compartment should be clean.
9) For quantitative analysis, make sure the absorbance/transmittance of the sample solution falls within the linear region of the standard curve. Dilute the sample as needed.

Problem Set

1   A solution of a light-absorbing substance has an absorption maximum at 540 nm. Is this ultraviolet, visible, or infrared light? Express the wavelength in Angstroms, microns, centimeters, and meters.

2 A $1.0\,\text{g}\,\text{l}^{-1}$ solution of compound X has a transmittance of 80% in a 1 cm cuvette at 500 nm. The molecular weight of X is $100\,\text{g}\,\text{mol}^{-1}$.
  a) Calculate the absorption coefficient of X.
  b) Calculate the transmittance of a $2.0\,\text{g}\,\text{l}^{-1}$ solution of X at 500 nm.
  c) Calculate the absorbance of both solutions (1.0 and $2.0\,\text{g}\,\text{l}^{-1}$).

References

1 Penner, M.H. (2017). Basic principles of spectroscopy. In: *Food Analysis*, 5e (ed. S.S. Nielsen), 79–88. New York, NY: Springer Science+Business Media.
2 Penner, M.H. (2017). Ultraviolet, visible, and fluorescence spectroscopy. In: *Food Analysis*, 5e (ed. S.S. Nielsen), 89–106. New York, NY: Springer Science+Business Media.
3 Segel, I.H. (1976). *Biochemical Calculations: How to Solve Mathematical Problems in General Biochemistry*, 2e. New York: Wiley. 441 p.

Answers to Problem Set

1 540 nm is in the visible range. $540\,\text{nm} = 5{,}400\,\text{Å} = 0.54\,\mu\text{m} = 5.4 \times 10^{-5}\,\text{cm} = 5.4 \times 10^{-7}\,\text{m}$.
2 a) $\varepsilon = 9.69\,\text{M}^{-1}\text{cm}^{-1}$.
  b) T of a $2.0\,\text{g}\,\text{l}^{-1}$ solution $= 64\%$.
  c) A of a $1.0\,\text{g}\,\text{l}^{-1}$ solution $= 0.0969$; A of a $2.0\,\text{g}\,\text{l}^{-1}$ solution $= 0.1938$.

Appendix V

Chromatography

What Is Chromatography?

Chromatography is a technique for separating components of mixtures as they are carried by a mobile fluid phase through a stationary solid or liquid phase. The separation is accomplished because the mobility of each solute depends on its distribution between the mobile phase and the stationary phase. Solutes with a higher affinity for the stationary phase will move more slowly than solutes with a higher affinity for the mobile phase.

The technique of chromatography was originated by the Russian botanist Michael Tswett in 1906. He used petroleum ether, a nonpolar solvent, to extract a mixture of green and yellow pigments from leaves. He then found he was able to separate these pigments by passing the extract through a column of fine powder. He called his method chromatography from the Greek meaning "color writing," because he used it to separate colored compounds. Today, chromatography is mainly used to separate colorless compounds, but the name remains to describe any technique based on the same principles.

Many advances have been made since chromatography was invented, and it is now one of the most widely used techniques in the fields of biochemistry, organic chemistry, and food chemistry. Chromatography may be used to separate subcellular fractions, specific substances in foods, and even closely related compounds such as simple sugars. Partition and adsorption chromatography separate solutes on the basis of solubility and adsorptive affinity, respectively; but other types can separate solutes on the basis of molecular weight (gel permeation chromatography) or ionic charge (ion exchange chromatography). All these techniques can be used, under properly controlled conditions, to identify solutes in solution. They may also be used to quantify the amounts of solute present.

The following is a brief overview of the principles and terminology of chromatography. Students are encouraged to see Ismail [1] for an excellent discussion of the basic principles and Reuhs [2] for a more thorough treatment on high-performance liquid chromatography (HPLC).

Chromatography Terminology

A rather specialized terminology has been adapted to describe and differentiate the various types of chromatography. Before describing the types of chromatography and the principles involved, it will be helpful to define a few terms.

Food Chemistry: A Laboratory Manual, Second Edition. Dennis D. Miller and C. K. Yeung.
© 2022 John Wiley & Sons, Inc. Published 2022 by John Wiley & Sons, Inc.
Companion website: www.wiley.com/go/Miller/foodchemistry2

Adsorbent. The stationary phase in adsorption chromatography. The stationary phase of partition chromatography may also be referred to as the adsorbent.

Development. The process of separating a solute from the sample using chromatography.

Eluant. The mobile phase once it is eluted from the column.

Elution. The process of removing a solute from the stationary phase.

Mobile phase. Material that is passed through the chromatography apparatus and stationary phase. It is a liquid in liquid chromatography and a gas in gas chromatography.

Normal phase chromatography. Chromatography with a polar stationary phase and a nonpolar mobile phase.

Origin. Physical location where the sample is first placed.

Reverse phase chromatography. Chromatography with a nonpolar stationary phase and a polar mobile phase.

Solvent or developer. Any mobile phase.

Sorbent. Any stationary phase.

Stationary phase. Material that is held in place during chromatography and that retains some component of the sample.

Support. The inert solid material serving to hold the liquid stationary phase in place in partition chromatography.

Types of Chromatography

The various types of chromatography are distinguished by the type of mobile or stationary phases used, the apparatus employed, or the basis of separation. Distinctions are not always clear-cut, and frequently one type of chromatography may have the characteristics of two or more other types.

The most basic distinction is based on the nature of the mobile phase. In gas chromatography, the mobile phase is a gas. In liquid chromatography, the mobile phase is a liquid. In gas chromatography, the sample is vaporized and carried by a carrier gas through a column packed with some material that acts as the stationary phase. Temperatures are usually quite high, and a specialized piece of equipment is required. In liquid chromatography, many different types of apparatus are used, ranging from a piece of filter paper suspended in a test tube containing some liquid solvent to highly sophisticated HPLC instruments that have precision pumps capable of forcing the mobile phase through the stationary phase at high pressures. Paper, thin layer, column chromatography, and HPLC all fall into the category of liquid chromatography.

The bases of separation of the components of a mixture by chromatographic means are the forces of interaction between solute molecules and the stationary and mobile phases. These forces may be ionic, polar, or dispersive in nature. Ionic forces result from the interaction of charged species (ions). Polar forces occur when polar molecules interact. Some molecules, such as water, have partially positive and partially negative regions as a result of electron withdrawing or donating groups. These polar molecules are permanent dipoles and are attracted to each other. Dispersive forces result when electrons in nonpolar or weakly polar molecules are displaced slightly by adjacent molecules. This causes the formation of a temporary dipole, which can then interact with other dipoles. These interactions are important when nonpolar solute molecules come in contact with nonpolar stationary phases. See Simpson [3] for a detailed description of these forces of interaction. Simpson [3] distinguishes among the various types of chromatographic methods based on the predominant type of molecular interaction involved. The following is a summary of Simpson's description of the various chromatographic modes for liquid chromatography.

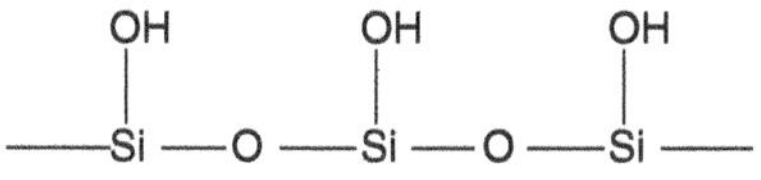

Figure V.1 Structure of silica gel, showing the hydroxyl groups on the surface.

Adsorption Chromatography (AC)

Adsorption is the adhesion of molecules to the surface of a solid. Silica gel is the most widely used adsorbent (stationary phase) in liquid adsorption chromatography. Alumina and diatomaceous earth are also used. The mobile phase in adsorption chromatography is usually a nonpolar liquid, such as heptane, but more polar liquids may also be used. Silica gel is a porous solid material with hydroxyl groups covering its surface (Figure V.1). Adsorption occurs when solute or solvent molecules interact with the hydroxyl groups. When water is present, most of the hydroxyl groups have water associated with them. Heating at 150–200 °C will remove the water, and heating is commonly used to activate silica gel before use. Water may interfere with the adsorption of solute molecules.

Liquid–Liquid Partition Chromatography (LLPC)

In this type of chromatography, a liquid is adsorbed on a solid matrix such as silica gel to form a liquid stationary phase. In LLPC, the stationary phase liquid is usually polar and the mobile phase liquid is nonpolar. Separation depends on the relative solubilities of solutes in the mobile and stationary phases. Solutes are partitioned between the two phases in the same way that solutes partition between immiscible liquids in a separatory funnel.

Bonded Phase Chromatography (BPC)

In BPC, the surface of the adsorbent is covered with organic groups, which are covalently bound to it. BPC is an improvement over LLPC, because the stationary phase can have many of the properties of a liquid but is more stable than when liquids are merely adsorbed to a solid matrix. Bonded phase columns are the most popular columns for use in HPLC. The most widely used bonded phases are nonpolar hydrocarbons containing 8–18 carbons per molecule. Polar and ionic organic groups may also be used. Silica gel is the solid support of choice. The organic constituents are bound to the silica gel by siloxane bridges (Figure V.2).

When the stationary phase is polar, a nonpolar mobile phase would be used. This is frequently called normal phase chromatography, presumably because it was the most common during the early history of chromatography. When the stationary phase is nonpolar and the mobile phase is polar, we have the so-called reverse phase chromatography.

Ion-Exchange Chromatography (IEC)

The stationary phase in IEC contains ionic groups (either cationic or anionic) that attract solute ions of opposite charge. Frequently, IEC is a form of BPC because the ionic groups are covalently linked to a solid matrix. The mobile phase in IEC is usually an aqueous buffer.

Gel Permeation Chromatography (GPC)

In GPC, solutes are separated on the basis of molecular size. The stationary phase in GPC is composed of tiny particles covered with pores of a defined size. Large molecules cannot penetrate the

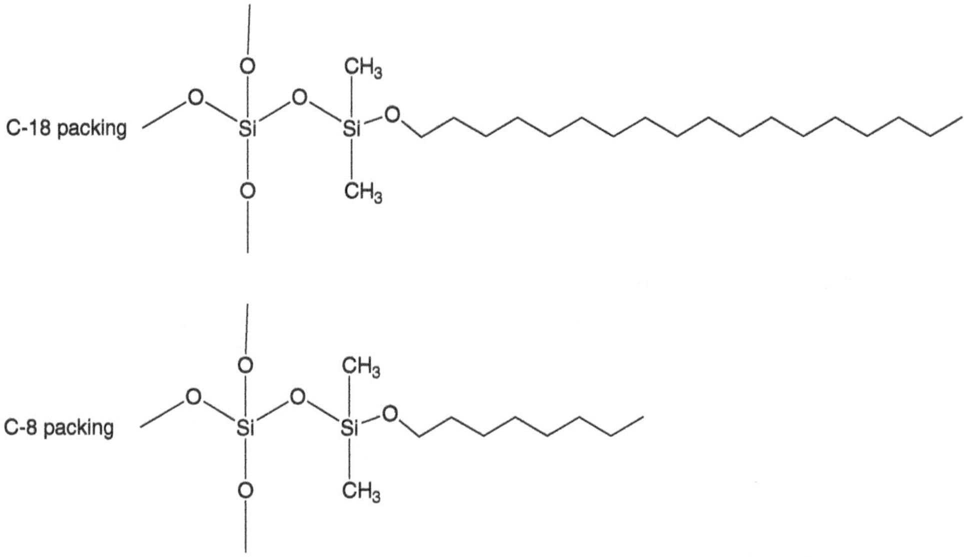

Figure V.2 Structures of stationary phase packing material with silica gel as the solid support. Note that the hydrocarbon chains are covalently linked to the silicon atoms in the silica gel. The C-18 and C-8 designations refer to the number of carbons in the hydrocarbon chains bound to the silica.

particles and move rapidly through the stationary phase. Small molecules enter the particles through the pores and are retarded. In GPC, solutes are eluted in order of decreasing molecular size. GPC is often referred to as size exclusion chromatography.

High-Performance Liquid Chromatography

HPLC is a specialized form of chromatography that has many advantages over more traditional chromatographic methods. It requires fairly sophisticated equipment specifically designed for rapid and efficient separation and quantitation of the components in a sample. It is widely used in the food industry. Some of its advantages include rapid analyses, specificity, high sensitivity (ability to detect low concentrations of the material in question), and simplified sample preparation.

Nearly anything found in food can be detected and quantified by HPLC, provided it can be solubilized in a suitable solvent. Sugars, organic acids, vitamins, amino acids, flavor compounds, additives, preservatives, and adulterants are examples of food components that have been detected and quantified by HPLC.

As the name implies, HPLC is a type of liquid chromatography in which a liquid mobile phase is passed through a stationary phase under high pressure. The stationary phase is contained within a narrow bore tube, called the column. It is common to classify HPLC as either normal phase or reverse phase (see Figure V.3). In normal phase chromatography, the stationary phase is a highly polar substance such as silica with amino or cyano groups bonded to it. The mobile phase is nonpolar, e.g. hexane or chloroform. In this system, polar substances are retained on the column longer than those that are less polar. Sugars, steroids, nitro compounds, amino acids, proteins, and peptides have all been analyzed with normal phase chromatography. In reverse phase chromatography, the stationary phase is nonpolar (typically a C8 or C18 hydrocarbon) and the mobile phase is polar (water, methanol, or acetonitrile). Nonpolar materials are retained on the column longer

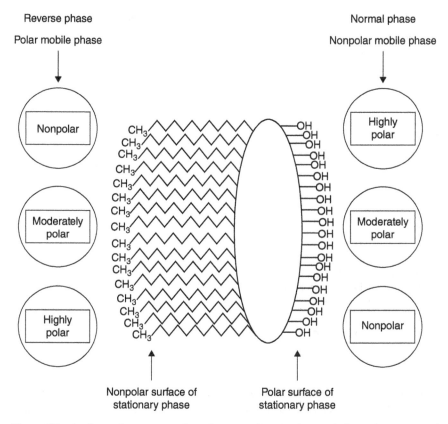

Figure V.3 A schematic representation of reverse phase and normal phase chromatography. The order of elution of sample components is shown in the rectangles, e.g. in reverse phase chromatography, highly polar components are eluted first. *Source:* Redrawn from [4].

in this form of HPLC. Amines, phenols, water-soluble vitamins, and essential oils are some of the materials that have been analyzed using reverse phase chromatography.

Two types of mobile phase delivery systems are common in HPLC systems: isocratic and gradient. In an isocratic system, the mobile phase remains the same and is pumped at the same flow rate for the entire separation run. This technique is suitable for many applications. In a gradient system, both the flow rate and mobile phase concentration change over the course of the separation. This is particularly helpful in methods development, where one can quickly run a scouting gradient to determine the best concentration at which to run the separation. Some methods require a stronger solvent toward the end of separation to elute particularly sticky compounds from the column.

The HPLC System

The actual configuration of an HPLC system depends on the types of samples being analyzed. However, most systems will have the following features:

1) A pump to deliver the mobile phase. The pump should be able to provide a uniform flow rate at pressure as high as 3,000 psi.

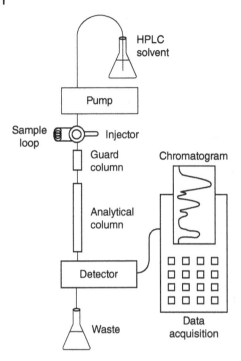

Figure V.4 A schematic representation of an HPLC system (Scott Langston, personal communication).

2) An injector to introduce the sample to be analyzed. Loop injectors are the most common type for routine analyses. A syringe is used to load the sample loop in the injector, and the mobile phase carries the sample toward the column.
3) A guard column to protect the more expensive analytical column from particulates and compounds that may irreversibly bind to the stationary phase. The guard column is essentially a smaller version of the analytical column.
4) An analytical column to perform the separation, based on the relative affinities of various components of the sample for the packing material.
5) A detector to detect the components of the sample as they elute from the column. A variable wavelength UV detector is the all-purpose detector of choice for most systems. More sophisticated and expensive systems may employ a mass spectrometer as the detector. LC-MS greatly enhances the capacity to identify components in a mixture.
6) An integrator/computer for data acquisition. The integrator/computer calculates the peak areas and heights and reports the retention times of the various peaks identified in the chromatogram.

A schematic of such a typical system is provided in Figure V.4.

References

1 Ismail, B.P. (2017). Basic principles of chromatography. In: *Food Analysis*, 5e (ed. S.S. Nielsen), 185–212. New York, NY: Springer Science+Business Media.
2 Reuhs, B.L. (2017). High performance liquid chromatography. In: *Food Analysis*, 5e (ed. S.S. Nielsen), 213–226. New York, NY: Springer Science+Business Media.

3 Simpson, C.F. (1982). Separation modes in HPLC. In: *HPLC in Food Analysis* (ed. R. Macrae), 79–121. London; New York: Academic Press. (Food science and technology).

4 Yost, R.W., Ettre, L.S., and Conlon, R.D. (1980). *Practical Liquid Chromatography: An Introduction.* Norwalk, CT: Perkin-Elmer. 255 p.

Suggested Reading

Lough, W.J. and Wainer, I.W. (eds.) (1995). *High Performance Liquid Chromatography: Fundamental Principles and Practice*, 1e. London; New York: Blackie Academic & Professional. 276 p.

Nollet, L.M.L. and Toldrá, F. (eds.) (2013). *Food Analysis by HPLC*, 3e. Boca Raton, FL: CRC Press. 1062 p.

Appendix VI

Electrophoresis

Introduction

Electrophoresis is a powerful technique that is widely used in food chemistry and biochemistry laboratories. Electrophoresis separates charged species on the basis of charge and molecular size.

The basic principle behind electrophoresis is that charged species move in an electric field. The direction and rate of movement of these species depend on the charge (+ or −), the number of charges, and the size of the ion. A cation (a positive ion) will migrate toward the cathode (the negative pole) and an anion (a negative ion) will migrate toward the anode (the positive pole). Usually, the species to be separated are applied to a solid support, normally a gel. Polyacrylamide gels are the most versatile although agarose and starch gels are also used. Paper and cellulose acetate strips are used as well. The big advantage of gels is that they can be prepared with varying pore sizes. This permits separations on the basis of size by a process called molecular sieving (think of the gel as a sieve through which small ions pass more readily than large ones).

The rate at which an ion moves through the supporting medium can be expressed as follows [1]:

$$\text{Mobility of a molecule} = \frac{\text{Voltage} \times \text{Net charge on the molecule}}{\text{Friction of the molecule}}$$

The voltage of the applied electric field can be set at any desired level. The charge on the molecule will be a function of the chemical nature of the molecule and the pH of the medium. The friction the molecule encounters will be a function of the size and shape of the molecule (larger molecules experience more friction because it is more difficult for them to move through the pores). Temperature and ionic strength may also affect ion mobility, so it is important to control them as well. It is always a good idea to include an internal standard when doing an electrophoretic analysis to correct for slight variations in conditions.

Polyacrylamide gels are prepared by mixing (in aqueous solution) acrylamide, bis-acrylamide, a catalyst (N,N,N',N'-tetramethylethylenediamine; TEMED), and a free-radical initiator (ammonium persulfate). The gels are formed as a result of the polymerization of acrylamide and N,N'-methylene-bis-acrylamide. This type of reaction should be familiar to you from your organic chemistry classes. The formation of polyethylene from ethylene and of polyvinyl chloride from vinyl chloride are similar reactions. All these reactions proceed via a free radical mechanism.

Food Chemistry: A Laboratory Manual, Second Edition. Dennis D. Miller and C. K. Yeung.
© 2022 John Wiley & Sons, Inc. Published 2022 by John Wiley & Sons, Inc.
Companion website: www.wiley.com/go/Miller/foodchemistry2

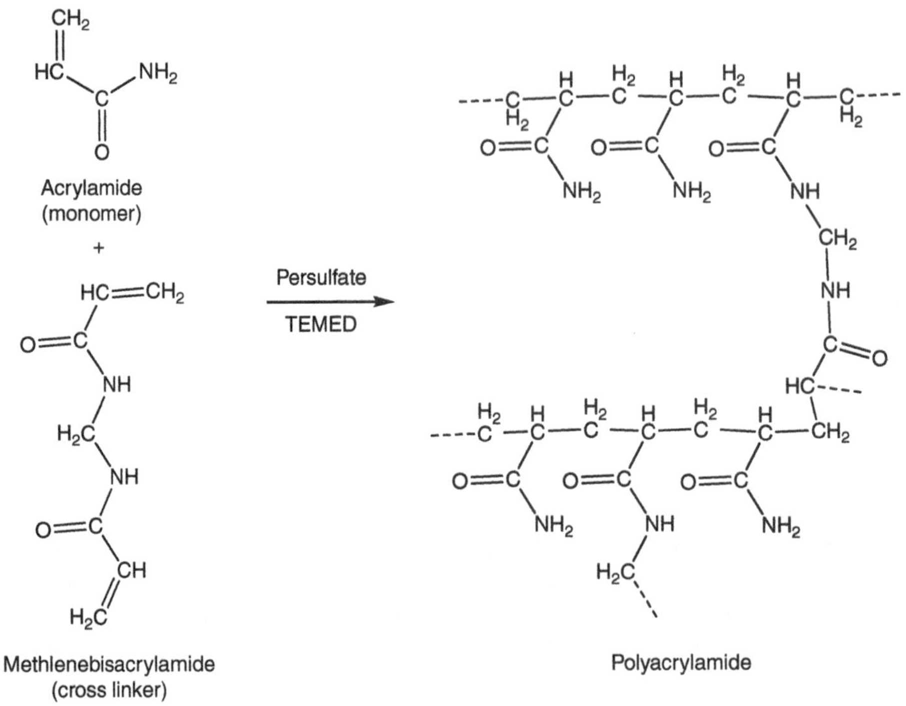

Figure VI.1 The formation of polyacrylamide from acrylamide and bis-acrylamide. Persulfate decomposes to sulfate free radicals and serves to initiate the free radical reaction. TEMED is a catalyst for the reaction. *Source:* Modified from [1].

Figure VI.1 shows the reaction for the formation of polyacrylamide gels. The pore size in the gel can be adjusted by changing the concentration of acrylamide in the mixture (for large pore sizes, use low acrylamide concentrations and vice versa).

The apparatus for conducting an electrophoresis experiment includes the following (Figure VI.2):

1) A buffer to hold the pH constant and to act as an electrolyte.
2) A buffer chamber.
3) A solid support (usually a gel).
4) A holder for the gel (either small tubes or a flat tray).
5) A power supply to generate an electric field.

Probably the most common application of electrophoresis is the separation, characterization, identification, and/or quantification of proteins. Proteins lend themselves well to electrophoresis because each protein has its own characteristic size and charge at a given pH. The charge on a protein is a function of its amino acid composition and the pH. At low pH, proteins generally carry a net positive charge because of protonation. At high pH values, proteins are negatively charged. At some intermediate pH (depending on the protein), proteins have no net charge and, therefore, would not migrate in an electric field.

To obtain maximum resolution of proteins, the following set of experimental variables must be optimized:

1) The buffer pH (maximal charge difference).
2) Gel pore size (maximal molecular sieving).

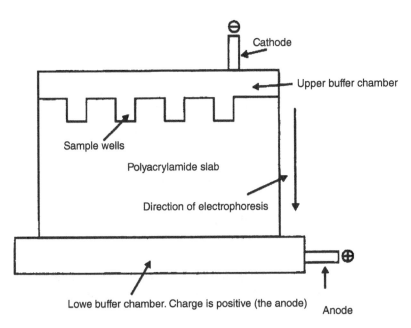

Figure VI.2 Gel electrophoresis apparatus. To operate, samples are applied to the wells, the apparatus is covered, and a current is applied. Negatively charged proteins migrate toward the anode.

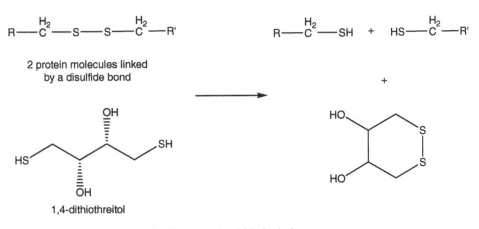

Figure VI.3 Cleavage of the disulfide bond by dithiothreitol.

3) Buffer ionic strength (sufficient for adequate buffering and current carrying capacity but low enough to allow large potential gradients).
4) Narrow starting zones (less total migration is required for complete zone separation when proteins are bunched together at the top of the gel).

Often, buffer systems that are designed to dissociate all proteins into their individual subunits are employed. Sodium dodecyl sulfate (SDS) (Figure 19.1) is an anionic detergent. It is the conjugate base of dodecyl hydrogen sulfate, a strong acid. SDS disrupts the quaternary structure of most multimeric proteins. Dithiothreitol and mercapto ethanol are reducing agents that cleave disulfide bonds in proteins (Figure VI.3). They may be used to dissociate polypeptide chains linked by disulfide bridges. Once the proteins are disrupted with these reagents, SDS binds to the various

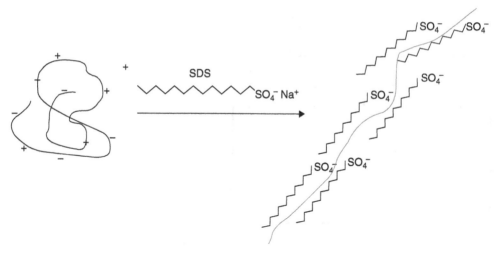

Figure VI.4 Binding of sodium dodecyl sulfate (SDS) to a protein molecule. The bound SDS anions force the conformation of the proteins into linear, rod-like conformations. *Source:* Redrawn from [2].

proteins in a constant weight ratio via hydrophobic interactions. The original charge on the protein now becomes insignificant compared to the negative charges provided by the bound detergent. It is estimated that proteins bind one SDS ion for every two amino acid residues. Thus SDS–protein complexes have essentially identical charge densities. Moreover, the bound SDS anions force folded up protein molecules into a linear, rod-like conformation since the negative charges repel one another and the conformation with the greatest separation of the charges is the linear conformation (Figure VI.4). Thus, the distance that proteins migrate in polyacrylamide gels is determined largely by the length of the molecule, which is strongly correlated with molecular weight (MW). A plot of the log_{10} of protein MW versus relative mobility shows a linear relationship between the two. Recall that relative mobility is expressed as the R_f value:

$$R_f = \frac{\text{Distance of protein band from the top of the gel}}{\text{Distance of tracking dye from the top of the gel}}$$

Thus a standard curve can be constructed for proteins of known MWs and used to estimate MWs of unknown proteins (Figure VI.5).

The MWs of unknown proteins are determined by comparing their migration rates with standards. Standards are a composite of several proteins of known MWs. Proteins of similar MWs will travel at the same rate and migrate the same distance within the gel. The absolute distance of migration will depend on the pore size in the gel – proteins travel farther when pore size is larger. As mentioned above, pore size can be adjusted by adjusting the concentration of acrylamide in the gel preparation. Gels made from a 11% acrylamide solution are recommended for separating proteins with MWs ranging from 14,000 to 70,000 daltons. For proteins in the MW range of 30,000–200,000 daltons, 7% gels are recommended [3]. A common standard contains the following proteins: myosin (MW = 200,000), β-galactosidase (MW = 116,250), phosphorylase B (MW = 97,400), serum albumin (MW = 66,200), ovalbumin (MW = 45,000), carbonic anhydrase (MW = 31,000), trypsin inhibitor (MW = 21,500), lysozyme (MW = 14,400),

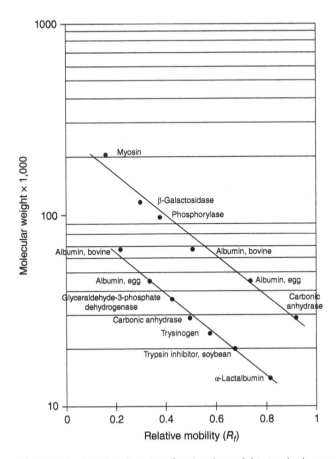

Figure VI.5 A calibration plot of molecular weight standards run on SDS PAGE. Note that the two parallel lines represent different electrophoresis runs on gels of different acrylamide concentrations. The upper line is from a gel with a lower acrylamide concentration compared with the lower line. *Source:* From [3].

and aprotinin (MW = 6,500). A different standard would be chosen when studying proteins with lower MWs.

See Smith [4] for more discussion on protein separation by electrophoresis and other techniques.

References

1 Switzer, R.L. and Garrity, L.F. (1999). *Experimental Biochemistry*, 3e. New York: W. H. Freeman and Co. 451 p.

2 Bio-Rad Laboratories, Inc (2020). A guide to polyacrylamide gel electrophoresis and detection [Internet]. https://www.bio-rad.com/webroot/web/pdf/lsr/literature/Bulletin_6040.pdf (accessed 6 May 2020).

3 Sigma Chemical Company (1998). SDS molecular weight markers in a discontinuous buffer. Technical Bulletin No. MWS-877L [Internet]. https://www.sigmaaldrich.com/content/dam/sigma-aldrich/docs/Sigma/General_Information/1/c2273inf.pdf (accessed 6 May 2020).

4 Smith, D.M. (2017). Protein separation and characterization procedures. In: *Food Analysis*, 5e (ed. S.S. Nielsen), 431–453. New York, NY: Springer Science+Business Media.

Suggested Reading

Garfin, D.E. (2009). One-dimensional gel electrophoresis. In: *Methods in Enzymology* (eds. R.R. Burgess and M.P. Deutscher), 497–513. Academic Press. (Guide to Protein Purification, 2nd Edition; vol. 463).

Wasserman, B.P. (1986). Detection of proteolysis by sodium dodecyl sulfate polyacrylamide gel electrophoresis: a demonstration of protein hydrolysis and electrophoresis fundamentals. *Journal of Food Biochemistry* 10 (2): 83–91.

Appendix VII

Glossary

Acid: A substance capable of donating a proton.

Aldonic acids: Organic acids formed when aldehyde groups on sugars are oxidized. Thus, the carboxylic acid group is on C-1 of the sugar. Examples are gluconic acid and galactonic acid, which form when glucose and galactose are oxidized.

Alginates: Polysaccharides composed primarily of mannuronic and guluronic acids joined by $1 \rightarrow 4$ linkages; occur naturally in brown seaweeds; used in salad dressings, pie fillings, structured fruit pieces, and many bakery and dairy products.

Anthocyanins: Plant pigments with colors ranging from red to blue (red in acidic conditions and blue in neutral to alkaline conditions). Anthocyanins are glycosides.

Anthocyanidins: Plant pigments that form when the glycosidic bond linking the sugar molecule to the main structure in anthocyanins is hydrolyzed, releasing the sugar molecule.

Antioxidant: A substance capable of slowing, stopping, or preventing oxidation.

Base: A substance capable of accepting a proton.

Buffers: Solutions containing a weak acid and its conjugate base or a weak base and its conjugate acid. Minimize changes in pH when small amounts of acid or base are added.

Carotenoids: A class of yellow, orange, and red pigments found in higher plants. Of the more than 600 known carotenoids, about 50 have vitamin A activity. Carotenoids are fat soluble and can scavenge free radicals and quench singlet oxygen.

Carrageenans: A family of sulfated linear polysaccharides that occur naturally in red seaweeds. They are polymers of D-galactose and 3,6-anhydro-D-galactose. Noted for their ability to form gels with milk.

Chlorophyll: Green pigments containing a porphyrin ring complexed with a magnesium ion.

Chromatography: A technique for separating components of mixtures as they are carried by a mobile fluid phase through a stationary solid or liquid phase.

Chromatograph: The apparatus used in the performance of chromatography.

Chromatogram: A visual record of the output from a chromatographic run. This could be visible colored bands appearing on a thin-layer sheet used in thin layer chromatography, a series of peaks on a computer screen showing the time-dependent elution of compounds in a mixture in high-performance liquid chromatography, or other forms of display of chromatographic results.

Electrophoresis: A technique for separating charged species (proteins and nucleic acids) on the basis of charge and molecular size.

Food Chemistry: A Laboratory Manual, Second Edition. Dennis D. Miller and C. K. Yeung.
© 2022 John Wiley & Sons, Inc. Published 2022 by John Wiley & Sons, Inc.
Companion website: www.wiley.com/go/Miller/foodchemistry2

Enzymatic browning: Browning that develops when cut or bruised surfaces of fruits, vegetables, and shellfish are exposed to air. Catalyzed by polyphenoloxidases.

Free radicals: Chemical species that contain one or more unpaired electrons. Most free radicals are unstable and highly reactive. Commonly identified by a dot signifying an unpaired electron, e.g., •OH is the formula for hydroxyl radical.

Galactomannans. Hydrocolloids with long backbone chains composed of repeating mannose units linked by β-1,4-glycosidic bonds and galactose side chains linked to mannose residues by α-1,6-glycosidic bonds. Common galactomannans include locust bean gum and guar gum.

Glyphosate: An effective herbicide that binds irreversibly to the active site of 5-enolpyruvyl-shikimate-3-phosphate synthase (EPSPS), thereby blocking the synthesis of aromatic amino acids in the plant.

Gums: Water-soluble polysaccharides capable of increasing viscosity or forming gels in aqueous systems. Examples include alginate, xanthan, guar, and carrageenan. Also called hydrocolloids.

Homolysis: The symmetrical cleavage of a covalent bond such that one electron of the pair forming the bond goes with each product. Homolysis of covalent bonds yields free radicals.

Hydrocolloids: Polymers that can be dissolved or dispersed in water and that produce thickening or gelling.

Hydrolysis: The breaking of covalent bonds through the reaction with water. Bonds susceptible to hydrolysis include peptide bonds, ester bonds, and glycosidic bonds.

Hydrolytic rancidity: Rancidity that develops when lipases attack triacyl glycerols that contain short-chain fatty acids. Most commonly occurs in products containing milk fat.

Hydroperoxide: A hydrogen-containing peroxide.

Leavening: A term derived from the Latin word *levo*, which means "raising or making light." Used in the baking industry to describe the rising of bread doughs and cake batters. The process of increasing the volume of batters and doughs by incorporating gas into them.

Lipid hydroperoxide: A hydroperoxide formed from lipids, usually polyunsaturated fatty acids.

Maillard reaction: A reaction between reducing sugars and amines resulting in browning and flavor development. A form of nonenzymatic browning.

Metmyoglobin: Myoglobin in which the complexed iron has been oxidized to the ferric (Fe^{3+}) form. Brown in color and accounts for the brown color of cooked meat.

Myoglobin: One of the major pigments in meat. A protein containing a heme group with complexed iron in the ferrous (Fe^{2+}) form. Dark purplish red in color.

Neutralizing value: The amount of $NaHCO_3$ that can be neutralized by 100 parts of a leavening acid.

Normal phase chromatography: Chromatography in which a polar stationary phase and a nonpolar mobile phase are used.

Oxidation: A reaction in which a chemical species either loses electrons, gains oxygen, or loses hydrogen.

Oxidative rancidity: Rancidity in foods that results from the oxidation of constituent lipids.

Oxymyoglobin: Myoglobin containing molecular oxygen bound to the iron in the heme complex. Bright red in color and contributes to the red color of fresh meat.

Peroxidases: A group of enzymes that catalyze oxidation–reduction reactions with peroxides as substrates. Peroxidases are ubiquitous in plant tissues and are among the most heat stable enzymes in plants.

Phytochemicals: Chemicals that occur naturally in plants.

Polyunsaturated fatty acids (PUFA): Fatty acids that contain more than one carbon–carbon double bond per molecule. Susceptible to oxidation because of the presence of the double bonds. Fats and oils from plant sources (soybean, corn, sunflower, peanut) tend to be high in PUFAs.

Reactive oxygen species (ROS): Oxygen-containing species that are more reactive than ground-state molecular oxygen (triplet oxygen). Examples include hydrogen peroxide, singlet oxygen, superoxide, hydroxyl radicals, peroxyl radicals, and alkoxyl radicals.

Reducing sugars: Sugars capable of reducing various oxidizing agents. Examples include glucose, fructose, lactose, and maltose.

Reduction: A reaction in which a chemical species either gains electrons, loses oxygen, or gains hydrogen.

Reverse phase chromatography: Chromatography in which a nonpolar stationary phase and a polar mobile phase are used.

Saturated fatty acids: Fatty acids with no carbon–carbon double bonds. Much less susceptible to oxidation than unsaturated fatty acids. Animal fats and coconut and palm oils are particularly rich in saturated fatty acids.

Singlet Oxygen ($^1O_2$): An excited (reactive) form of molecular oxygen. Contains an empty π*2p orbital and thus is an electrophile. Reacts readily with carbon–carbon double bonds in PUFA to form lipid hydroperoxides.

Superoxide ($O_2^{\bullet-}$): A free radical formed when molecular oxygen gains one electron. Produced by phagocytic cells to kill bacteria and viruses. It also forms when electrons leak from the electron transport chain and combine with oxygen.

Surfactants: Amphiphilic molecules containing one or more nonpolar, as well as polar regions, thus serving as emulsifying agents at the interface between oil and water.

Triplet oxygen ($^3O_2$): Ground-state molecular oxygen. The predominant form of oxygen in the air. Relatively unreactive toward most organic molecules at physiologic temperatures unless reactions are enzyme catalyzed. Not considered a ROS.

Uronic acids: Organic acids formed from the oxidation of $-CH_2OH$ groups on sugars without oxidizing the aldehyde groups. Examples include glucuronic and galacturonic acids.

Vitamin E: A fat-soluble vitamin. Functions as an antioxidant to protect cell membranes from oxidation. Eight naturally occurring compounds (four tocopherols and four tocotrienols) have vitamin E activity. α-Tocopherol is the most active form.

Warmed over flavor (WOF): The off-flavor that develops in cooked meat when it is refrigerated and then reheated.

Xanthan gum: An extracellular polysaccharide produced by the bacterium *Xanthomonas campestris*. Highly soluble in hot and cold water and imparts high viscosity at low concentrations. Used as a thickening and stabilizing agent in syrups, salad dressings, and gravies.